THE URBAN LIFE AND URBAN LANDSCAPE SERIES

Suburban Steel

THE MAGNIFICENT FAILURE OF THE LUSTRON CORPORATION, 1945–1951

Douglas Knerr

The Ohio State University Press
Columbus

Library of Congress Cataloging-in-Publication Data

Knerr, Douglas, 1960–.
Suburban steel : the magnificent failure of the Lustron Corporation, 1945-1951 / Douglas Knerr.
p. cm. — (The urban life and urban landscape series)
Includes bibliographical references and index.
ISBN 0-8142-0961-0 (Cloth : alk. paper) — ISBN 0-8142-9031-0 (CD-ROM)

1. Prefabricated houses—United States—History. 2. Lustron Corporation—History. 3. Housing—United States—History—20th century. 4. Business failures—United States. I. Title. II. Series.
TH4819.P7 K59 2004
338.7'69081'0973—dc22

2003025440

Cover design by Dan O'Dair
Type set in Perpetua
Printed by Thomson-Shore, Inc.

The paper used in this publication meets the minimum requirements of the American National Standard for Information Sciences—Permanence of Paper for Printed Library Materials. ANSI Z39–48-1992.

9 8 7 6 5 4 3 2 1

CONTENTS

LIST OF ILLUSTRATIONS

PREFACE

"What's that weird house doing there?" my aunt asked my father as we drove past a Lustron during the summer of 1975. That's when I first heard the story, and although I'm sure I feigned interest in the back seat of the car as my father told us how these unusual houses came to be, I did remember the story. Although many aunts have no doubt subsequently sniffed at the idea of living in a porcelain-enameled steel prefabricated house, I no longer feign interest in the story. *Suburban Steel* is a culmination of that childhood memory and an attempt to recount a portion of the history of a company that in the late-1940s claimed it had "the house America was waiting for."

In Mansfield, Ohio, where I grew up, there was one Lustron on my street and one on the next street over, which was a bit strange for a community comprised mainly of brick colonials and sprawling ranches. My father's extensive knowledge of business history and his familiarity with Lustron kept me wondering what might have been. He always thought that the houses were well conceived and that the company could have revolutionized the housing industry. After I moved off to school, we still discussed how the houses in our neighborhood were faring—one great, the other not so well, although at this writing they are both in fine shape once again and occupied by families that appreciate their significance.

I'd like to tell you that the Lustron house and the company that created it made a crucial contribution to alleviating the enduring shortage of affordable single-family housing in the United States during the post–World War II era. It made an impact but was ultimately unsuccessful in its quest to apply innovative production technologies to the mass housing market. And that's unfortunate, since technological advances in the housing industry have been slow to evolve and have been less than effectively used to expand opportunities for affordable home ownership across the income spectrum. The 2,000 or so Lustrons that remain of the 2,500 produced in the late 1940s are thus relegated to living museum status, physical relics of a road not taken and an opportunity unrealized. They are magnificent artifacts of a time when a confident postwar society set its sights on the housing crisis in much the same way as it approached war production. Indeed, outproducing the Axis powers in some ways proved an easier task than revolutionizing the entrenched economic, political, and social elements of the American housing system.

Scattered primarily throughout the Midwest, Lustron houses have been the focus of considerable media attention in recent years. An Emmy-winning PBS documentary, stories on cable television home shows, numerous websites, and homeowners' conventions have kept Lustron in the public eye. People are still attracted to the novelty of a house made of porcelain-enameled steel that looked like a typical ranch house but contained numerous time and space-saving features to facilitate "modern living." Lustrons were also one of the few "houses of the future" that actually got beyond the drafting table, but it was for a future that never arrived. This makes them enduringly interesting, especially to those like me who find the history of "yesterday's tomorrows" particularly intriguing.

I've been in many Lustrons over the years, but one encounter during my graduate school years focused my interest on this project. While writing a history of a Cincinnati company that owned the Chicago Vitreous Enamel Products Company, the technical progenitor of Lustron, I met a number of people who were present at Lustron's genesis. Several employees at ChiVit, as it was then called, vividly recounted the Lustron era and the excitement over the prospect of contributing to and profiting from a new type of housing production. A casual conversation turned into a dissertation and, finally, to this book.

The journey was neither expeditious nor without frequent detours, and I have accumulated many debts along the way. My professors in the history department of the University of Cincinnati, especially Zane Miller and Gene Lewis, provided valuable guidance at the inception of the project and throughout the dissertation process. As a visiting faculty member at Northern Kentucky University, I was fortunate to tap the expertise of colleagues Mike Adams, Frederic Krome, and Eric Jackson. In my current job in the Evelyn T. Stone University College at Roosevelt University, I am indebted to the university and college administration for supporting this and a number of other forthcoming projects. Special thanks go to my faculty colleagues Mike Bryson, Mike Maly, Jack Metzgar, and Dan Headrick.

Additional thanks are due Zane Miller in his role as editor of the Urban Life and Landscape Series at the Ohio State University Press. I'm sure Zane thought he was rid of me at the hooding, but his consistent support and thoughtful direction throughout the revision process once again demonstrated that he has few peers in the field of urban history and fewer still as teacher, mentor, and friend. I also extend my gratitude to Heather Lee Miller of the Ohio State University Press for providing valuable assistance during the revision and production process, to peer reviewers Kristin Szylvian and Carolyn Loeb, as well as to other members of the Press staff who assisted in the production of this book.

My parents have my enormous gratitude for their consistent encouragement and support throughout the years, and my wife deserves special thanks for countless kindnesses initiated and returned over our many happy years together. This book is for my family—immediate, extended, past, present, and future.

INTRODUCTION

Foundations

"A house is a machine for living."
—Le Corbusier, *Vers une architecture*

The Lustron Corporation was once the largest and most completely industrialized housing company in the history of the United States. Beginning in 1947, Lustron manufactured porcelain-enameled steel houses in a one-million-square-foot plant on 106 acres of land in Columbus, Ohio. At peak production the plant used more electricity than the entire city of Columbus. Just one year after its incorporation, Lustron's capital investment exceeded that of the next forty largest housing firms. With nearly $40 million in federal funds and support from the highest levels of government, the company planned to produce 100 houses per day, each neatly arranged on specially designed tractor-trailers for delivery throughout the nation. Though radical in its use of porcelain-enameled steel, the Lustron house—a one-story, gabled-roof ranch with a bay window and side porch—looked much like other postwar-era dwellings. Behind its traditional facade, however, lay hopes and expectations for a new era in American housing.

The unprecedented scale and scope of the company's operations attracted intense scrutiny. Years of interest and anticipation regarding the industrialization of the house construction process focused attention on Lustron's strategy, structure, problems, and prospects. At last, the nation would see if a large, highly capitalized firm could usher in a new age of affordable mass-produced housing. The efficiencies of continuous production, integrated manufacturing, and economies of scale promised to lead the American housing industry away from its decentralized, undercapitalized, and inefficient past toward a level of rationalization and organization found in most other sectors of the industrial economy. As Senator Ralph Flanders of Vermont observed in 1947, "if Lustron doesn't work, let us forever quit talking about the mass produced house."[1]

Indeed Lustron "didn't work," and in 1951 collapsed amidst foreclosure and bankruptcy proceedings. The company had produced only 2,498 houses. Lustron's brief and often tumultuous existence reflected the expectations, frustrations, triumphs, and follies of American society in the immediate postwar era. Fresh on the heels of a victory won as much on the production line as on the battlefield, Lustron's plans to revolutionize the housing industry seemed well within the context of American assumptions about industrial progress and the promise of a prosperous future. Two of the company's national advertising slogans—"The House America Has Been Waiting For" and "Lustron: A New Standard for Living"—effectively expressed the nation's expectations for accelerated progress and for the idea of a house as a "just reward" for a successful fight to defend the American way of life. As historian John Morton Blum observed, "the house and all things that went into it, 'the American home,' best symbolized of all things material a brave new world of worldly goods."[2]

While millions of American families realized their dreams of home ownership in the decades after the war, Lustron was able to fulfill only a small fraction of those dreams. Lustron's founder, Carl Strandlund, intended to become the Henry Ford of housing, yet ended up most often compared with the failed automotive entrepreneur Preston Tucker. Like the Tucker Torpedo, the Lustron house was born of postwar optimism and incorporated innovations that foreshadowed future trends. Lustron was far more significant, however. The company's failure marked a watershed in the history of the American housing industry. Although people did not "quit talking" about industrialized housing, enthusiasm for its role in the transformation of the housing industry at large markedly waned.

The Lustron saga is interwoven amidst a number of significant threads of the American experience and represents a logical extension of technological, organizational, and architectural trends evident in the housing industry since the late-nineteenth century. Lustron's product—the single-family house—holds a central place in our culture and over the years has been as much a focus for intellectual construction as in the physical sense. From the beginning Americans have consistently wrestled with how best to think about housing as well as how to build the best house. The nation's industrial maturity throughout the nineteenth century altered long held conceptions of house, home, and community, forcing issues surrounding the transformation of housing forms and settings to the forefront of social concern. Since then, some of the best and brightest minds in the nation have considered how house, home, and community could be reconstructed in light of a fundamentally changed society. The evolution of the philosophy and practice of American housing is consequently a very messy story, full of economic, political, and social entanglements woven deeply into our institutional fabric as well as into our psyches.

Disentangling the threads is no easy task. The average single-family house represents a complex web of relationships and transactions beginning with the moment of its conception and potentially extending beyond its physical existence. The housing industry has both reflected and fostered this complexity. A patchwork of small-, medium-, and large-scale enterprise operating under countless local variations and employing a range of modern and ancient methods and materials, it has seemingly defied the forces of industrial progress for more than a century. For those with the means, it offers a level of product customization based on individual tastes and expression that few industries can match. For those without the means to engage it at its best, it has been far less responsive. However, in the period leading up to Lustron's genesis in the immediate post–World War II period, social, economic, and political forces coalesced to open a number of intriguing possible alternatives for those Americans who found decent housing an elusive prospect.

Lustron's "magnificent failure" reveals frustrations as well as opportunities created by the housing industry's distinctive evolutionary path. The realization in the late-nineteenth and early-twentieth centuries that the housing industry would not follow a typical pattern of rationalization, consolidation, and technical innovation in service of the mass market generated widespread concern in both the public and private sectors. Considerable intellectual and practical talent was marshaled to address a series of housing crises that initially focused on the working classes and eventually expanded to include a broad segment of the American population. Frustrations with an industry so disorganized and unresponsive to the housing needs of a modern industrial society gave rise to a reform impulse that initiated an important national discussion about housing and eventually drew the federal government into the equation.

Like the industry itself, governmental involvement was often neither rational nor well organized. A patchwork of engagement and disengagement, cooperation and compromise, rhetoric and genuine reform, federal housing policy both provided opportunities and raised obstacles for the housing industry. The relationship is particularly intriguing, for it demonstrates the interdependency of the public and private sectors as policy makers and capitalists attempted to create a coherent structure that would enable greater numbers of Americans access to the best that the housing industry could offer. Lustron was one place where public and private interests intersected in the late 1940s, and it remains an excellent example of the complex character of interactions between public policy and private industry in the pursuit of social improvement. The relationship paralleled in microcosm the range of emotions and actions evident in the national debate over housing policy and practice in the decades prior to Lustron's founding.

Reconciling a deep-seated belief in the sanctity of private property, free enterprise, and the ideal of home ownership with the reality of a housing system that failed to extend the benefits of decent shelter to millions of American families has confounded generations of the nation's best thinkers. The intricate character of product, process, and setting has made comprehensive reform difficult. Indeed, it is hard to ask a simple question about housing. Questions such as "Where do you live?" or "Where do you want to live?" can contain volumes of meaning for both the questioner and the respondent. In the late 1940s the Lustron Corporation asked America where it lived and whether it would like to live in a new kind of house, one made of porcelain-enameled steel and produced in one of the largest factories in the nation. Millions answered yes—but only several thousand actually live in one today. What happened? To answer that question with any satisfaction requires some entanglement in economic, political, and social discussions surrounding the development of the American housing system during the first half of the twentieth century.

An examination of the physical provenance of Lustron's product, particularly interest in applying industrial methods to its construction and deployment, provides an opportunity to unite the multifaceted dimensions of housing reform with political and market realities. A number of possible technical and productive paths lay before the housing industry in the early decades of the century, one of which was the manufacture of an entire house in a factory setting. Factory-made housing involved shifting as many elements of housing construction as possible from the building site to the factory.[3] The benefits of this approach seemed to hold the answer to a number of vexing problems. Although conventional builders sometimes employed systematic production techniques in the construction process, the house itself was still largely "manufactured" on-site. Even the "mail order" houses found in the catalogs of Sears, Roebuck & Co. and Montgomery Ward in the early-twentieth century required conventional fabrication and assembly methods to construct the house on a lot. In contrast, the factory approach implied a total restructuring of the American housing industry. Applying economies of scale to housing production meant increased affordability, better quality through precision engineering and manufacture, continual production regardless of weather conditions, and the flexibility to introduce new innovations and amenities in a short period of time.[4]

The idea of factory-produced housing also satisfied frustrations with the industry's seemingly disorganized and irrational structure and held the promise of aligning the industry with the productive ideal of other "modern" industries. The prospect of houses produced as readily and affordably as automobiles was a seductive vision, and it attracted proponents and evangelists from far and wide. For architects the allure was based in part on the intellectual and creative challenges inherent in understanding the dimensions of a complex integrated system with transformational social potential. But it was not merely the dream of a small segment of the architectural profession. The industrialization of housing captured the imagination of engineers, manufacturers large and small, industrial designers, construction firms, real estate developers, venture capitalists, labor leaders, government policy makers, newspaper editors, and the American public. Widespread calls for a factory solution for "the industry that capitalism forgot" became an early-twentieth-century corollary to the hand wringing of the post–Apollo era when many Americans asked If we can put a man on the moon, why can't we meet other social, political, and economic challenges? It seemed unbelievable that Americans could not find a way to apply sophisticated mass production methods to the housing industry to create, in Henry Ford's words, "a productive organization that delivers in quantities a useful commodity of standard material, workmanship and design at a minimum cost."[5]

Yet a productive realignment of the housing industry was far more complicated than first imagined. Such a transformation assumed major technological advances and a shift in complex economic relationships in the housing market. The decentralized character of the industry made any type of unifying vision extremely difficult to conceptualize, let alone implement. Diverse and parochial, insular and hidebound, the industry was nevertheless politically, socially, and economically connected to American life in fundamental ways. Indeed, the nature of its structure had defined and reflected particularly American values throughout its existence. This provided a tremendous reserve of practical power and philosophical protection to forestall those who would attempt to impose an alternative vision for the industry on any grounds other than its own economic self-interest.

However, the appeal of factory-produced housing on grounds of economic efficiency and political expediency was undeniable in the early part of the century. It fit well within the progressive impulse for rationalization and systematization in economic and social life. Early in the century, architects and manufacturers viewed prefabrication as a method to improve the efficiency of the traditional house construction process by supplying standardized components at a reduced cost. As the technology matured, the industry could then begin to compete with conventional construction by offering affordable and higher-quality houses to the "entry level" market segment. Finally, building on manufacturing and marketing savvy, prefabricated housing firms could eventually reinvent the entire housing industry by rationalizing the production of houses just as American business had done in other economic sectors.[6]

To a federal government increasingly drawn into housing reform by the inability of the housing industry to serve the needs of an expanding industrial society, mass production of housing in a factory setting seemed worthy of serious consideration as a policy objective. The allure of "dramatic" technology, especially one based on rational organization, was understandable. It was a high-profile idea, easily explained via the automotive analogy, and it seemed so wonderfully modern. The factory was a powerful metaphor for modern life and a key to the promise of a more prosperous future based on increased material consumption. Just as factories provided new and exciting products for Americans to display in and around their houses, why not subject the house itself to the same consumptive paradigm?

Although the process of housing construction seemed amenable to industrialization, the setting proved far less adaptable. Concern over setting—how a house fits into its physical and social surroundings—emerged in the early part of the century as a key component of reform ideology. Housing reformers came to embrace the community development model, an approach that emphasized the totality of community, including streets, parks, shopping, religious life, and additional social amenities as essential elements for a suitable living environment. The locus and form of the communitarian ideal were inherently suburban, and in the early decades of the twentieth century, economic forces, federal policy, and social influences coalesced to encourage the expansion of suburban development and identified the ownership of a single-family house in a suburban setting as an American ideal.

Although suburbs had existed nearly as long as cities themselves, the prospect of uniting the benefits of the suburban lifestyle with an expanding middle-class mass market offered an intriguing solution to America's housing dilemma. As the federal government became more involved in housing reform after the start of World War I, it pursued policies that established suburbia as the preferred locus of all things American. Despite significant direct involvement during the war, however, the return to normalcy in the 1920s meant that it would be left to the private housing market to populate suburbia with affordable housing. When this did not materialize, interest in an industrialized approach increased. While it was easy to envisage the mass production of well-conceived and affordable housing for the mass market, it was unclear how such ventures would be financed and organized, which materials would best lend themselves to mass production, how consumers would finance the purchase of a factory-made house, and how such housing could be suitably situated in communities, as well as hundreds of other unknown questions and considerations.

By the time Lustron emerged on the scene, the housing industry had taken significant steps toward industrialization, though its progress was not readily apparent to the average American homebuyer. The industry had not been static, but it had yet to achieve the productive potential and resulting affordability necessary to serve a budding suburban mass market. Aided by public and private initiative, industrial advancement had progressed along several parallel fronts. A steady convergence of technologies and processes was underway, and as the evolution progressed the federal government grew increasingly interested in a production-oriented solution to the nation's persistent housing problem.

The industrialization of housing component production represented the "hidden rationality" of the industry. From the mass production of nails and glass to the increasingly sophisticated prefabrication of moldings, doors, staircases, wall units, and plumbing stacks, efficiencies and cost savings inherent in this type of productive refinement benefited the entire housing industry. Additionally, the use of industrialized processes at the construction site became increasingly prevalent. Site prefabrication, as it came to be known, applied organized systems of production to an entire housing development. Creating an "outdoor factory" where teams of workers focused on one task across the site promised to increase production, reduce labor costs, and enable builders to purchase materials in large quantities. Site prefabrication provided a powerful model of industrialization that held great potential in serving a mass market with quality housing and affordable cost.

Yet even site prefabrication seemed too provincial an approach to the industrialization of the American housing system. Certain that the housing industry should follow the automotive industry's path of centralized development, proponents of factory-built housing argued that only large, highly capitalized firms with sufficient leverage over raw-material costs and tremendous productive capacity could provide decent and affordable single-family housing to the American mass market. The nation needed high volume production from coast to coast, and only factory-built housing could solve a pressing social need by bringing organization and rationality to the field of residential construction. Beginning with a small chorus of voices in the early-twentieth century, calls for a productive realignment of the American housing system increased dramatically in the following decades, finally to reach a crescendo in the post–World War II era.

The triple crises of depression, war, and the veterans' housing shortage of the immediate postwar era provided political legitimacy and economic refuge for the idea of factory-produced housing. These disruptions exposed the inefficiencies and shortcomings of the traditional housing industry for all to see and led Americans for the first time to call for an activist federal role in seeking alternatives to the status quo. From restructuring the home finance credit system, to seeding the private construction market through government entities such as the Reconstruction Finance Corporation, to building communities for low-income citizens or housing for defense workers, federal involvement in the housing market during the 1930s and 1940s shook the housing industry to its core and provided a window of opportunity for alternative approaches to residential construction. Included among the range of responses of the federal government was an increasing interest in industrialized housing and related technologies. Throughout the 1930s and especially during the war, federal housing programs supported the growth and development of what was popularly known as the prefabricated housing industry.

A desire for stability and predictability in the midst of social and political chaos perhaps made the idea of prefabrication more palatable. In the search for an "architecture of order," the prefabricated system provided a logical paradigm that affirmed the power of industry, the advancement of technology, and the democratic benefits of market capitalism. Initially a loose conglomeration of enterprising individuals, startup firms, and grant-funded researchers, prefabrication moved from what one observer termed "a state of mind" in the early 1930s to a "movement" by the end of the decade that had garnered significant commercial and governmental interest. The energies expended on prefabrication were indeed impressive. Architect and educator John Ely Burchard commented on the scope of effort: "Here is a galaxy of well-known names; here are the fruits of incalculable hours of thought and research by able men; here are ideas that cover in principle almost everything that a human being might conceive in the field of redesign of house structure."[7] The well-known names who tacked prefabrication during the 1930s included eminent architects such as Wright, Gropius, Le Corbusier, Neutra, and Fuller; prominent industrialists such as Owen Young, Foster Gunnison, and Howard Fisher; corporations such as U.S. Steel, Johns-Manville, General Electric, and Celotex; nonprofit organizations such as the Russell Sage Foundation and the John B. Pierce Foundation; universities such as Purdue and the Massachusetts Institute of Technology; and government agencies such as the U.S. Forest Products Laboratories, the National Bureau of Standards, the Federal Housing Administration, and the Tennessee Valley Authority.

The combination of crisis and reaction in government and the prospect of commercialization presented an unprecedented opportunity for an alignment of interest between government and industry. Through depression and war, interest in prefabrication in particular and the industrialization of housing construction in general moved gradually to a more central position within federal housing policy. After the war, the acceptance of the idea that production technologies could solve the perpetual American housing crisis as surely as it won the war is well reflected in the federal government's support of the Lustron Corporation as an exemplar of the factory-built approach. The concentration of such a large amount of support and anticipation around Lustron makes it a useful vantage point from which to assess the interplay of technological and social forces, the influence of economic and political processes, and particularly the role of the federal government in the search for housing solutions.

Interest and support of housing technologies during this period seemed to represent the brief emergence of a "middle tier" of federal housing policy. Between the indirect but essential support for the housing industry through the home finance credit system on one hand, and direct involvement in the construction of housing on the other, funding the development of production-oriented housing technologies represented an alternate path. Neither of the two tiers proved entirely satisfactory. Traditional builders welcomed indirect support but bristled at the idea of government competition—even if the government intended to serve markets they had abandoned. On the other hand, housing reformers welcomed direct federal involvement in community building, especially for low-income Americans, but realized that the government credit guarantees still left a decent house beyond the reach of a great many American families.

Government support for industrialized housing might at once provide incentives for private industry but direct its development in ways beneficial to all citizens. Improved productive technologies might bridge the housing gap by lowering the total costs of housing in all market segments. This addressed better housing issues on a wholesale level and held the promise of gradually lessening the federal government's direct role in the housing industry by improving the private sector's ability to serve broad markets. Federal support for new housing technologies promised to reduce total housing costs for entire classes and groups of citizens further down the economic scale. This represented a course somewhere between commercial and social Keynesianism and assumed that the incubation of promising housing firms would advance specific public policy interests as well as generally encourage the expansion of one of America's core industries.

With prefabrication economically nurtured and politically legitimized, support for a factory-produced housing venture such as the Lustron Corporation seemed a logical approach to meet the challenge of an acute housing shortage following the demobilization of the armed forces during 1945 and 1946. Optimism surrounding prefabrication appeared well placed. Prefabricated housing firms built approximately 200,000 units during World War II—nearly all, however, were demountable, or impermanent, structures for government housing projects adjacent to defense plants and installations. Although far from ideal, the industry had shown promise, and many articles in business and architectural publications during the war predicted a rosy future as the industry developed more sophisticated products and more efficient production processes. The technological allure of prefabricated housing laid the foundation for its prospective economic and political benefits. With the promise of continued federal assistance in obtaining capital, raw material, and markets, a revolution in housing appeared imminent.[8]

Lustron first intersected with federal housing policy as the Truman administration wrestled with a shortage of "decent" housing for veterans returning from World War II. Veterans' organizations led a politically powerful lobby that resonated throughout the country and put tremendous pressure on the federal government. The administration responded with the Veterans' Emergency Housing Act (VEHA), which provided numerous financial incentives for manufacturers of prefabricated housing. Born out of the aims of the VEHA and stoked with capital from the Reconstruction Finance Corporation (RFC), Lustron ultimately received nearly $40 million in government loans to pursue the production of factory-made houses—far more federal assistance than any other prefabricated housing firm of the era.

On one level, the willingness of the federal government to support Lustron reflected a continuation of wartime industrial policy. The veterans' housing crisis was fundamentally a production issue, and the government possessed many effective, war-proven mechanisms to spur industrial production. However, while there was little doubt that the economy would eventually accomplish the task, the political pressures and social forces of the immediate postwar era heightened interest in more dramatic technological solutions. The cause of homeless veterans mobilized a wide range of political interests whose representatives descended on the Truman administration with significant force. The connection between home ownership for America's heroes and the maintenance of the American way of life was clear. The media trumpeted the cause with great vigor and even raised the specter of homeless veterans agitating for social upheaval as in Europe after the Great War. While policy makers did not anticipate a revolution, they felt intense pressure to respond to the crisis and quickly embraced an approach that combined "socialistic" elements of direct federal aid for housing but tempered within the context of wartime industrial planning.

But who would receive the economic bounty of crisis? Conventional builders were too slow and wedded to an old-fashioned production system. The attraction to prefabricated housing nurtured during the war now turned more serious and set the stage for a new partnership. Lustron's appearance reflects a combination of technological innovation and engineering expertise driven to the marketplace by entrepreneurial and organizational talent. It also provides a window into an area of entrepreneurship that relies on an activist government to create opportunities for industries, firms, and individuals in the furtherance of public policy goals. Historians in recent years have demonstrated the powerful and often overlooked significance of this type of commercial relationship.[9] Successful partnerships between government and industry to serve national goals became a fundamental part of federal strategies to lift the nation out of the Great Depression and win the Second World War. Infrastructure improvements such as dam and road building during the depression, and the unprecedented alliance between government and industry in the prosecution of the war, evidenced the power and potential of government entrepreneurship.

The housing industry entered into partnerships with the government to serve the need for housing in a number of New Deal programs, but when the federal government assumed a major role in providing housing to support an unprecedented internal migration that accompanied military mobilization, interest in housing technologies increased dramatically. The benefits of government entrepreneurship resonated within the prefabricated housing industry. The industry hoped that government capital, market guarantees, and other forms of federal largesse would transform prefabrication from a peripheral commercial development into a major force in the rationalization of the housing industry.

Lustron's origins reflect an alignment of interests between a company with an innovative housing technology and the emerging policy goal of providing housing to a greater number of Americans by promoting the productive capacity of the nation. As industrialist Henry Kaiser capitalized on the federal government's infrastructure and defense needs during war and depression, so Lustron intended to capitalize on the acute postwar housing crisis. In 1946 and 1947, providing affordable housing to returning veterans became the goal to which Lustron attached its commercial aspirations. Though reticent to publicly criticize efforts to provide housing for veterans, many policy makers questioned whether the nascent prefabricated housing industry could deliver on its promise of mass-produced housing without extraordinary levels of government support. In times of economic and political crisis, federal support for firms building infrastructure or the weapons of war seemed logical and necessary. Yet how far could the relationship be extended?

Connections between the public and the private sector characteristic of government entrepreneurship often prove mutually beneficial, especially when accepted by the public as a reasonable strategy for the implementation of desirable policy goals. For example, government loans to business through the Reconstruction Finance Corporation had aided economic recovery during the depression and financed a significant portion of America's war production effort. However, the question remained whether the nation would accept the same structure in peacetime to fight the "war" for affordable housing, especially for returning veterans. The experience of the Lustron Corporation in particular and the prefabricated housing industry in general reflects the boundaries of midcentury government entrepreneurship. In contrast to the defense and infrastructure industries, connections to the public interest proved less concrete and far more susceptible to controversy when a more general social need such as housing became the focus of governmental efforts.

Unlike conventional construction where builders could enter the market with little up-front capital, prefabricators needed a large initial investment for plant and equipment as well as for financing and distribution systems. Private investors frequently balked at providing the capital necessary to realize the benefits of mass production. The infancy of the industry, the scale and complexity of the production process, and uncertainties surrounding marketing and distribution discouraged many venture capitalists. Further, fears that new technologies, methods, and materials might destroy much of the value of an initial investment in one type of production system accentuated financial barriers to entry.

Yet the reluctance of private investors failed to discourage supporters of prefabrication in the Truman administration. The limits of government entrepreneurship in the social realm fostered a lively debate within the administration and in society at large. The controversy mirrored the larger tensions surrounding continued governmental activism and centralized economic control that undergirded broad political divisions in postwar society. Supporters argued that the successful production and marketing of factory-made houses deserved federal support because it served a desirable social goal. The reluctance of risk-sensitive private capital markets to provide sufficient resources to this capital-intensive industry demanded that the government step in and fill the gap. Advocates claimed that governmental support of Lustron would "jump start" the entire industry, which could then turn to the private market for its subsequent capital requirements. It might also further the cause of industrialization across the entire housing industry and thus benefit all Americans.

Opponents of federal support for the prefabricated housing industry immediately expressed a number of reservations. Foremost, they questioned the wisdom of providing large amounts of public funds to an unproven industry and argued that federal capital allocation to industry during peacetime undermined the free market and stifled entrepreneurship. While the RFC had served a useful purpose during depression and war, its continued existence promoted a "creeping socialism" in American life. In their view, support for prefabricated housing, especially the large amount of government capital invested in Lustron, reflected a de facto government "takeover" of the industry. Lifting the mantle of free competition, conventional builders and smaller producers of prefabricated housing fervently objected to high-level governmental support for Lustron. Many asked how any other firm could compete against Lustron when the government had subsidized its huge productive capacity and national marketing campaign. These groups also expressed fears that the government would protect its investment in the company by granting it allocations of scarce materials or favoring the company for military and defense housing contracts.

The ongoing controversy and constant pressure of the political spotlight compressed Lustron's window of opportunity and heightened expectations for immediate success. The company made many mistakes and miscalculations typical of a new firm in a young industry, yet none proved so damaging as the belief that the government would maintain its support until the company outgrew its initial difficulties. As criticism mounted and political pressure became too great, the government abandoned Lustron as quickly as it had embraced it. Unable to compensate for the withdrawal of government backing, the company soon collapsed. This study argues that Lustron failed because it exceeded the acceptable limits of government sponsorship during the immediate postwar era by extending the business-government partnership to serve a peacetime social need, a goal the American public ultimately failed to support.

Lustron, then, was a stillborn enterprise. It began with the goal of providing affordable single-family dwellings for wage earners—the mass market for mass production. Initially unable to provide a house that most wage earners could afford, critics of government policy were quick to declare the company a failure. Lustron survived the embryonic stage of development only to be abandoned as it was discovering how to access the market segment that rationalized large-scale manufacture. Its association with the administrative problems and political entanglements surrounding the RFC hastened its demise. Ironically, by 1950 Lustron had proposed creative solutions to many of its startup problems and stood ready to offer a new range of houses for a broad market segment. But the window of opportunity had disappeared in the face of mounting political pressures and the ability of conventional builders to increase production. In the wake of Lustron's failure, the government's interest in a middle-tier policy based on technological and commercial cultivation of industrialized housing all but disappeared.

The web of political controversy enveloping Lustron tended to accentuate its mistakes and mask its substantial accomplishments. For a brief moment, the company reflected the can-do optimism and spirit of accomplishment of wartime America extended into the postwar future. It appeared at the right time with a ripe idea—and with an engaging person to lead the revolution. Carl Strandlund, Lustron's founder and president, was a brilliant engineer with numerous patents to his name and awards for innovations in war production. Strandlund worked for the Chicago Vitreous Enamel Products Company, a leading manufacturer of enameling products and a pioneer in the use of porcelain-enameled steel panels for commercial construction. Mr. Strandlund went to Washington in 1946 in search of steel, a tightly controlled commodity, to facilitate expansion of his firm's enameled-steel building division. Officials at the Civilian Production Administration, the peacetime successor to the War Production Board, literally laughed at his request. However, when Strandlund suggested that the same technology could be used to mass-produce houses, government interest rapidly increased.

Strandlund seemed well suited for the role of government entrepreneur. He was in a unique position to introduce a promising idea with a sound technological base into the entrepreneurial climate created by the federal government's interest in housing production. Although initially a political neophyte, Strandlund learned quickly and used his dynamic personality to elicit support from the administration, Congress, veterans' organizations, and, perhaps most significantly, the national press. Strandlund possessed traits that made successful government entrepreneurs: the ability to work outside formal bureaucracy, a strong "public friendly" personality, and the skill to cultivate positive media attention. The linkage of personal skills, technological feasibility, and public policy goals put the Lustron Corporation in business.

The company faced the same challenges of manufacturing and marketing that had confounded earlier attempts at mass production of affordable housing. However, Lustron made significant progress toward the alleviation of a number of traditional barriers facing the industry. Factors of scale and scope alone demonstrate the company's significance to the history of the American prefabricated housing industry. The scale of Lustron's operations far exceeded any previous, or subsequent, prefabricated housing venture. At a target production level of 100 houses per day, the company would easily have been the largest producer of housing in the nation. In 1947 only 854,600 single-family (nonfarm) dwellings were produced.[10] That one firm could potentially provide such a large percentage of the nation's housing output indeed evidenced the dawn of a new era in the housing industry.

Lustron's large capital resources and resulting financial power also represented a significant departure from previous patterns of development in the prefabricated housing industry. Before Lustron, most firms were either small, undercapitalized, entrepreneurial ventures, or peripheral divisions of large companies that sought to develop housing applications for their core products. Contemporary observers most frequently cited undercapitalization as the major barrier to success in the industry. Albert Farwell Bemis, an industrialist, author, and prominent evangelist for prefabrication during the 1930s, noted that large capital investment in mass-production technology had revolutionized all major industries and would certainly do the same for housing.[11]

Lustron's significant financial resources, like those of other large manufacturers, enabled it to "start big" and exploit all the advantages that size brings to the marketplace. The use of steel as a construction material also facilitated the attainment of economies of scale. Lustron introduced mass-production machinery such as presses, punches, welding apparatus, and enameling ovens to the housing industry. The company's application of proven commercial building technologies to the production of single-family dwellings also represented a key element of its competitive strategy. The technology and manufacturing expertise employed by Lustron evolved from Chicago Vitreous's production of porcelain-enameled "art deco" building facades popular in the 1930s, and from the porcelain-enameled steel architectural system developed for the construction of gasoline service stations.

Lustron's commitment to integrated manufacturing and large investment in production technology reflected its desire to implement a "countrywide production line," where rationalization and organization of the house construction process extended from raw material handling to the final processes of on-site assembly. The company introduced a number of innovations to achieve that goal. Lustron's factory was a model of efficiency and automation and took full advantage of advancements in the machine tool and material handling industries during World War II. With the assistance of a prominent industrial engineering firm, Lustron created what *Fortune* magazine called "a marvel of American manufacturing ingenuity."[12] Continuous-process steel fabrication, multilevel welding operations, and ovens capable of processing 12,000 square feet of enameled steel per hour enabled a synchronization of production that reduced the need for warehousing raw materials. In fact, Lustron's operation was much like the just-in-time inventory control strategy currently employed by many manufacturers. Lustron also integrated its distribution system into the production process. With the assistance of a transportation firm, the company designed a new type of tractor-trailer that transported the completed house from the factory to the erection site. In the last phase of production, the trailer became part of the material handling process as completed parts moved from various assembly lines directly onto the trailers in reverse order of their need at the assembly site.

Another important aspect of Lustron's production strategy was its relationship with organized labor. Represented by the American Federation of Labor (AFL), the building trades had long been suspicious of prefabrication because it appeared to threaten local autonomy and control of the house construction process. From the beginning, however, Lustron cultivated a close relationship with the national leadership of the AFL. Consequently, the company and the AFL negotiated a comprehensive labor pact that guaranteed the exclusive use of union labor at the factory and at erection sites. The Lustron/AFL agreement became a model for the prefabricated housing industry and demonstrated labor's willingness to support new approaches to housing construction.

Lustron's strategies to meet the challenges of producing factory-made houses represented only one element of success. Technological innovation and manufacturing expertise put mass production within reach, but did little to address the impact of prefabrication beyond the shipping dock of the factory. The relationship between firms and their social environment, particularly in the interplay of technology and the marketplace, is aptly demonstrated by the experience of the Lustron Corporation and the prefabricated housing industry.[13] The technologies that promised to solve the housing crisis and rationalize a "backward" industry proved incredibly alluring, yet they raised many issues that spoke directly to American assumptions about housing—assumptions nurtured by decades of tradition.

As the industry moved from component prefabrication toward the production of an entire house in a factory, the process gradually began to separate the house from its environment. A singular focus on prefabrication as an industrial process within a factory precluded consideration of its effects on such key issues as marketing, distribution, financing, public acceptance, and the placement of prefabricated houses in neighborhood and community settings. Integrating both the process and the consequences of prefabrication proved a difficult task, for it revealed a key tension between the need for standardization to accommodate mass production and the forces impeding standardization in the marketplace. Lustron's responses to the challenges created by mass production and mass marketing of its house provide insight into the interplay of economic and social forces.

The company's recognition of and relation to the major postwar trend of suburban development demonstrates tensions that frequently arise between technological innovation and its subsequent social application. From the outset, Lustron envisioned its house within a suburban context. The company focused considerable attention on the integration of its marketing plans with emerging patterns of suburban development. Lustron's recognition of the importance of environmental factors, such as site and neighborhood planning, reveal a commitment to address issues beyond the mere process of prefabrication. Indeed, the company devised detailed plans for the "proper" placement of its houses in neighborhood and community settings.

Proper community development demanded that the economic and social needs of residents figure prominently in the planning process. Without these elements residents would be susceptible to social disorganization and neighborhood deterioration. Federal housing strategies were deeply influenced by the communitarian approach. The desire to provide affordable single-family dwellings in suitably designed neighborhood physical and social environments is reflected in the Housing Act of 1949, which expressed the goal of providing "a decent home and a suitable living environment for every American family." The reference to "a decent home and a suitable living environment" in the Act demonstrated the acceptance of the idea of community as an essential component of good housing policy. However, the desire to make *all* housing community housing raised an important issue for manufacturers of prefabricated houses. By separating the house from its environment, prefabrication represented a potential threat to the communitarian ideal.[14]

Indeed, throughout its history the prefabricated housing industry had demonstrated a profound disregard for environmental issues and focused primarily on the process of manufacture and technological or design innovations. The "F.O.B." pricing policies of the industry offer telling evidence of its lack of concern beyond the shipping dock of the factory. F.O.B. is a common abbreviation of "free on board," a purchasing arrangement in which the customer assumes the cost of shipping a product from the factory to the delivery site. In essence, once the sale had been made, the industry considered its task complete. Urban planners and architects advocating the community development model frequently criticized the prefabricated housing industry for its "process" focus. The *Architectural Record* cautioned that firms that "expect to reach the top by inventing the one perfect standardized panel are foredoomed to bitter disappointment. . . . People everlastingly buy neighborhoods, not mere houses. Houses dumped *en masse* with no further responsibility . . . can only foul the nest." And *The American City* observed that "the picture of an area crowded with . . . factory-produced houses . . . is not a pleasant one to contemplate."[15]

The industry faced a key problem of integrating its product into prevailing assumptions about housing and the community ideal. Lustron's response to the community issue encompassed a number of strategies. The company developed the Planning Guide, which outlined specific instructions for the proper integration of the house into community environments. The introduction to the Guide reflected a holistic view by asserting that the company's product was "essentially a house, lot, and its environment." Under the slogan "good planning is good Lustron business," the Guide offered a view of the ideal "Lustron community."[16]

The Planning Guide reflected the influence of what Marc A. Weiss has called the "community builders"—the creators of the modern residential subdivision. Weiss identified a "revolution in community building by the 1940s" that witnessed the merging of communitarian impulses with the rise of suburban development. Subdividers became "private planners" who created and implemented concepts and forms that "came to be accepted as good planning." These included street design and layout, cul-de-sacs, planting strips, consideration of topography, the arrangement of the house on the lot, setback lines and separation parameters, and the integration of community elements such as recreational facilities, schools, and shopping centers.[17] Lustron's Planning Guide addressed each of these topics in great detail, and the company credited the Federal Housing Administration's 1940 publication *Successful Subdivisions* as the basis for the Guide's suggestions, the same publication now quoted by historians as evidence of how the community building approach was "fully incorporated as public values to be standardized and emulated."[18]

Lustron's association with community building was a significant step for a manufacturer of factory-made housing. Community building came easier for firms incorporating only certain elements of prefabrication in the building process. William Levitt and Sons, for example, used "site prefabrication," a process that took advantage of economies of scale in raw material purchase and employed an effective system of labor rotation. Under Levitt's system, the house remained a part of its environment from the beginning of the construction process.[19] A manufacturer of factory-made housing, though, faced greater difficulties in assuring a community setting.

Lustron established a fleet sales department to facilitate the creation of Lustron communities and to encourage large suburban developers to use Lustron houses. The company cultivated a close relationship with American Community Builders, Inc., the developers of Park Forest, Illinois, an archetypal postwar subdivision.[20] Lustron also marketed its community approach to suburban developers in most major cities in the United States. This became a major goal of the company's sales strategy, which defined the creation of a "complete package"—lot, house, and community—as a key to sustained long-term growth.

The alternative to the community marketing strategy presented a number of serious difficulties, especially for a manufacturer of a porcelain-enameled steel house, even though the Lustron house had a traditional ranch-style design. Marketing a "radical" house without exerting control over its placement meant that the house would undoubtedly come into conflict with "conventional" housing in established neighborhoods. Moreover, the Lustron house would face the various obstacles by which settled neighborhoods enforced their particular vision of planning and community design, such as building codes, deed restrictions, and zoning ordinances. Prefabricated housing had traditionally met substantial opposition from these elements of urban control because the public often perceived prefabrication as cheap, impermanent, and acceptable only in an atmosphere of crisis, such as defense housing during World War II.[21] Lustron's community marketing strategy, along with the traditional design of the house itself, sought to overcome negative images of prefabrication. Further, the creation of new Lustron developments in suburban or semirural areas enabled the company to "solve" the problem of possible restrictions on prefabricated houses.

However, economic pressures imposed by manufacturing bottlenecks and engineering delays forced the company to postpone full implementation of the community marketing strategy until late 1949. Up to that point Lustron was more concerned with generating sufficient sales to keep afloat until it could develop connections and expertise in the community-building field. Consequently, the company marketed the house solely through a franchise dealer network, which by definition relinquished some control over environmental issues to individual dealers. The franchise contract established standards for placement of the house, but during Lustron's start-up phase the administrative structure to enforce the standards was not fully implemented.[22]

Lustron's marketing approach focused on the image of the house itself in an effort to overcome negative public perceptions of prefabricated housing. The company's advertising stressed the design continuity of the Lustron house with conventionally built houses and pointed out the superior performance and durability found in a dwelling made of porcelain-enameled steel. Lustron tried hard to convince potential buyers that the inflexibility of the house, specifically the difficulties in adding on new rooms or changing interior floor plans, was more than offset by its technological and architectural benefits. Indeed, customer surveys revealed few complaints about the lack of opportunity for individual expression and indicated that owners were willing to forego flexibility for other advantages, especially low maintenance costs.

While public acceptance of the Lustron house was encouraging, acceptance by local governments and mortgage lenders proved more elusive. Lustron encountered difficulties integrating its house into settled neighborhoods. Building codes presented a significant and enduring barrier. Thousands of different state and local building codes closely governed the fabrication of individual house components but failed to consider the performance of the components. For example, even though Lustron's porcelain-enameled steel wall panel outperformed conventional "stick built" walls in terms of strength and durability, it failed to meet code standards and was therefore prohibited in many jurisdictions. Confronting the code problem, Lustron urged dealers to submit plans and specifications of the house to local officials at the earliest possible date and provided supporting documentation, which included examples of favorable treatment in other jurisdictions and citations of successful legal appeals involving the Lustron house. The company also participated in a national lobbying campaign for code reform through a trade association for manufacturers of prefabricated housing. The campaign urged the adoption of uniform, performance-based codes that did not judge prefabricated houses within the context of traditional materials and construction technologies. Lustron also aided dealers on the local level to obtain code variances and provided a "press packet" to alert local and national media about the need to reform building codes in light of "advanced technologies of housing construction."

The company also faced the difficult problem of financing the purchase of a factory-made house. Financial institutions considered such a house chattel property until it was affixed to a lot and proved reluctant to provide mortgages until the erection process was complete. Securing approval by the Federal Housing Administration (FHA) was also crucial in order to make FHA-insured mortgages available to homebuyers. However, since FHA standards were not uniform throughout the nation, wide variations existed in the requirements of FHA and mortgage lenders—requirements that frequently bore no relationship to the quality of the house or the safety of the loan.[23] Another problem was bridging the gap in financing between the time the house "package" left the factory and the time the mortgage financing of the finished house was arranged for the ultimate consumer.

Lustron attempted to confront these substantial barriers by establishing a new type of financing plan in partnership with a large mortgage-lending firm. The company also envisioned the creation of a "Lustron Acceptance Corporation" modeled on the process of consumer financing popularized by the automobile industry. The company's efforts to change the nature of financing in the prefabricated housing industry attracted widespread attention and praise in the financial press, but attempts to effect a revolution in financing inevitably faced resistance. The prospect of mass-produced housing raised concerns among mortgage lenders, who feared that technological improvements, style innovations, and the prospect of lower-cost homes would threaten the value of existing houses on which they held mortgages. Since mortgage lenders controlled funds for a credit-intensive industry, their perceptions of the impact of factory-made housing—right or wrong—carried substantial influence. As General Motors' chairman Alfred P. Sloan observed, "the political (and economic) power of the pressure groups that would be antagonized by such a revolutionary approach to one of our greatest industries might well make any organization hesitate, and justly so."[24]

Yet Lustron did not hesitate. Access to substantial government capital and the ebullient optimism of its founder fostered a "full speed ahead" culture that at times served the company well and at times did substantial damage. Indeed, Lustron's responses to longstanding problems within the prefabricated housing industry produced a number of innovative approaches to the production and marketing of factory-made houses. However, no amount of innovation could solve the problems created by its intimate relationship with the federal government. Political controversy stalked the firm throughout its existence and produced intense scrutiny from friends and foes alike. Though partially due to the unprecedented scale of operations and the uniqueness of the house itself, national interest in Lustron was consistently driven by the level of federal involvement in the venture.

Economist Ludwig von Mises observed that businesses must frequently fail for the market to gain useful information. Lustron taught the nation that the technology for the mass production of affordable housing was well within reach. It demonstrated that American homebuyers would live in prefabricated houses during nonemergency conditions and that prefabrication could be integrated into traditional neighborhood and community settings. It revealed the complexities of marketing factory-made housing and highlighted tensions between technological advances and their applications in the culture- and tradition-bound housing industry. And it raised important issues in the realm of business–government relations by testing the boundaries of institutional and firm interaction in the realm of social policy.

Reflecting on the Lustron experiment, one of the company's original dealers observed in 1960 that "the Lustron Home was the forerunner of what will be the future of the mass-production home market. The home industry moved out of the 'Model T' stage for the short period of Lustron's existence and demonstrated that high quality factory-made homes could gain widespread acceptance."[25] Original purchasers consistently praised the durability, ease of maintenance, and attractive design features of the house. Subsequent generations of Lustron owners reported a similar level of satisfaction in a survey completed during the 1980s. Years after the demise of the company, the emotions that the Lustron house evokes in contemporary observers remain strong. As the following chapters reveal, the emotional content of the Lustron experiment in general—from unbridled optimism to inconsolable despair—confirm the ongoing appeal of this magnificent but ultimately unsuccessful attempt to bring the American housing industry into the modern age.

CHAPTER ONE

An Uncomfortable Convergence: Housing Reform and Industrial Influences

In love of home, the love of country has its rise.
—Charles Dickens, *The Old Curiosity Shop*

The Lustron Corporation was not the first to challenge the status quo of housing in America. It had been going on at least since the Second Great Awakening of the 1830s and 1840s when evangelicals and "sanitarians" linked public health issues to housing and identified a host of social ills associated with inferior housing. These reformers advanced notions of "decent" housing and identified the evils of substandard, crowded tenements in the core of America's developing industrial cities. Substandard housing also often encompassed conditions of tenure such as lodging, boarding, or other temporary states of residence—all of which early reformers believed fostered individual and societal instability.

As definitions and descriptions of inferior housing emerged, at the other end of the spectrum an ideal for individual and societal stability appeared in the form of the single-family house on the urban periphery. As industrial cities developed through midcentury and a new middle class gained prominence, society increasingly celebrated ownership of this type of housing as virtuous and ennobling. Its iconic status reflected the Lockean tradition of the sanctity of private property fused with the American ideal of individualistic achievement. The power of those influences, coupled with an expanding industrial economy, fueled the conquest of the suburban frontier. The original suburbanites—proprietors, managers, professionals, and others who performed the "mental work" of the new economy—slowly but surely made their way to the more distant fringe communities, shattering the bonds of the old "walking city" and forever changing the social character of American life. Aided by transportation innovations, industrial cities became increasingly segregated by social and economic class and politically transformed by a new and powerful suburban-based socioeconomic constituency.[1]

The idyllic appeal of the suburban neighborhood, with its well-appointed houses, paved streets, sanitary sewers, and generous lots soon captured the imagination of the American public. However, the dream of suburban home ownership remained elusive for the majority of American families. Many decades would pass before the nation would accept that every family had a right to that aspiration or agree to support systems that provided both the productive and financial resources necessary to make such dreams possible. Yet a chorus of voices in the late-nineteenth century known as the progressives began to lay the foundation for a restructuring of the American housing system.

The Progress of Reform

As the twin engines of urbanization and industrialization remade the social, political, economic, and physical landscape of American society in the late-nineteenth century, the plight of the urban working classes impelled reformers to political action and challenged them to adopt broader conceptions of the role of housing in society. The language, attitudes, perceptions, and strategic approaches of the housing reform movement that began in the late-nineteenth century remained significant, clarifying as well as occasionally obscuring subsequent approaches to the perpetual American housing crisis.

Progressive reformers focused intense energy on housing issues during the heyday of the movement from the 1890s through the Great War. Progressive housing reform drew on many sources of expertise and tapped the talents of social activists, architects, industrialists, academics, journalists, bureaucrats, and politicians. Housing fascinated the progressives since many believed that environmental factors exerted a powerful influence over character and behavior. This belief gave rise to new definitions, new approaches, and new possibilities. Properly conceived and implemented, housing could order a disordered society, encourage democracy in the face of undemocratic influences, and foster morality, virtue, and character to counter myriad threats presented by "decadent" urban lifestyles.[2]

Reformers embraced a number of strategies, each embodying the spirit of the movement. Initial efforts focused on the tenements. Tenement reform encompassed a variety of commercial and governmental strategies to achieve change. "Model tenement" associations appeared in New York City at midcentury and quickly spread to other industrial cities. Associations often employed a strategy known as

"Philanthropy and 5 Percent," which involved forming a commercial entity to construct model housing for workers and limiting investor dividends to keep rents affordable. The effort to control capitalism for the benefit of the deserving poor was an innovative idea, although in many cases the impulse to control extended to officious moral oversight by "pantry snooping" reformers anxious to impose their vision of the virtuous life upon residents.[3] While the approach attracted a great deal of attention and produced some successful results, it had little overall impact on America's tenement problem.

Progressive beliefs in the power of government encouraged reformers to lobby for legislation and advocate sociological jurisprudence to effect changes in housing. Arguing that the state's power to protect and foster public safety and health extended to housing, progressives believed that government should play a key role in reform. New York City again led the way by enacting a series of tenement house laws from the late 1860s through the turn of the century that restricted the type and placement of buildings, mandated fire escapes and other safety features, and allowed for the razing of existing structures in the interest of public health. Other cities adopted the New York model of tenement reform to varying degrees. Although these laws did not alleviate the tenement house (urbanist Lewis Mumford observed that "housing reform by itself has only standardized the tenement"), the appearance of housing legislation on the local level was a significant step. Appealing to issues of public safety and child endangerment, progressive reformers prepared a seat at the reform table for government and convinced at least some Americans that governmental action was necessary to mitigate the social costs of an ill-housed populace.[4]

Another significant progressive era development in housing reform was the professionally planned company town. Though not a new idea—the first American company town dated to 1645—the manner in which progressive capitalists worked with housing reformers to develop new conceptions of worker housing was influential and reflective of broader trends and themes. The impulse to control the workforce was a hallmark of the "welfare capitalism" movement of the late nineteenth century. Savvy firms and their owners understood that constructing an environment to house their workers and initiating various controls over their social, economic, and political behavior was a natural extension of the processes of rationalization, standardization, and systemization occurring throughout the industrial economy. Applying cost/benefit and systems analysis to worker housing proved just as revealing as the application of these analytical tools to other facets of the business enterprise. As surely as firms used horizontal and vertical integration to gain control over raw materials and markets, so too they understood that the company town could be a valuable tactic to control their labor costs, forestall unionization, and shape a labor force best suited to the demands of the modern industrial enterprise. Planned company towns thus reflected the larger dynamic processes of industrial development well underway by the late nineteenth century and typified progressive desires to control and rationalize those processes.[5]

The company town movement attracted a range of interest well beyond the field of industrial relations. Housing reformers, representatives of the "professional welfare" community, architects, city planners, and other design professionals vied for commissions or attempted to exert intellectual influence over the conception and implementation of company towns. Drawn by the opportunity to design a planned community from a blank slate and embracing the challenges of constructing a total living environment based on the latest thinking across a number of disciplines, these professionals advanced ideas and strategies that would influence the larger field of housing reform.[6]

Progressive social ideology manifested itself in many aspects of the company town, and by the early-twentieth century architects and planners grew more sophisticated in their approach. New conceptions about housing the working classes had been percolating throughout the movement, and professionals working on company towns found a ready laboratory in which to experiment with new approaches. By the middle teens, the housing reform movement was in the midst of a shift from a "shelter model," where reform focused on improvement of buildings such as tenement houses, to a "community model," which emphasized the planning and control of numerous environmental variables that influenced the lives of residents. This was a significant change, for the community approach framed the discussions about housing reform in both urban and suburban contexts for decades to come. Subsequent attempts to alleviate America's housing problem needed to address the issue of community environment in one way or another to be considered viable.[7]

Beginning with well-known examples of the planned company town such as Pullman, Illinois, and extending into the more sophisticated "second generation" of company towns built during the teens and twenties, the community development model held sway over the cadre of professionals who planned, landscaped, decorated, and supervised daily life in company towns across the nation. In the "play space" of the company town, where conventional restrictions of the outside world were mitigated and where corporate financing provided a range of possibilities, designers could practice social engineering freely to foster both the "efficiency and uplift" of the working residents as well as satisfy the aesthetic and economic requirements of their corporate clients. In a positive sense, the company town forced designers to view housing as a dynamic process and enabled them to experiment with new ideas in a controlled environment.[8]

For progressive housing reformers forced to consider society at large, however, things were more complicated. The harsh economic realities of the free market—

the decentralization and inefficiency of the housing industry, the increasing class and social stratification of the suburbanizing industrial city, and the prevailing view that government had at best a limited role in housing reform—each presented daunting challenges. The "community organization problem" implied a vast array of interdependent reform issues centered on the concept of decent housing. In the coming decades, reformers wrestled with every conceivable issue as the tabula rasa of the American urban hinterland was reshaped by the rise of modern suburbia. The company town looked simple by comparison, for the reform constituencies faced innumerable problems and an opposition who employed powerful philosophical, historical, economic, political, and social arguments against the extension of the opportunity for decent housing to a greater number of Americans.

Yet by the eve of America's entry into World War I, progressive reformers could reflect upon a number of significant accomplishments. Although disparities remained—vast numbers of urban working-class families remained in urban slums while increasing numbers of middle-class families set their sights on the good life in the emerging suburban fringe—reformers had succeeded in initiating and focusing a broad national discussion about the viability of America's housing system. The discussion engaged creative and energetic minds from a number of professional fields, from the public and the private sectors, and from a variety of political and economic backgrounds. Far more than simply identifying a nationwide housing problem, progressives defined the problem in a broad social context. Their thinking evolved from a localized focus on the tenement house to a more sophisticated view of housing and its connection to dynamic social, economic, and political processes. They began with ideas about improving shelter and arrived at concepts of "sheltering communities," a logical progression given the complex underpinnings of the built environment and the movement's own focus on environmental determinism.

Progressives also established a baseline definition for "decent" housing, argued that all Americans had a right to its benefits, and highlighted roles that the public and private sectors might play in the search for reform strategies and tactics. The identification of housing reform as a nationally prominent issue that demanded the attention of both government and private industry made it increasingly difficult for either to ignore. Although crisis conditions motivated each sector, from the time of U.S. intervention in World War I the interplay of public and private interests in the struggle for housing reform presents a useful framework for analysis. It explains at least in part how a public/private "hybrid" corporation such as Lustron could emerge with so much fanfare in the late 1940s with millions of government dollars seeding its dream of extending affordable housing to a mass market.

The relationship between housing reform and the housing industry is complex and multifaceted—at times contentious, at times mutually beneficial, and frequently mischaracterized. Historians are far more familiar with the evolution of public responses to the housing crises of the twentieth century and have constructed a coherent interpretive arc that chronicles the increasing involvement of the state in housing reform. Yet the role of business and industry in this story has not received prominent attention and thus remains uncomfortably seated within the overall picture. The housing industry has served as a convenient foil against which "heroic" reformers have railed. It has correctly been blamed for the perpetuation of economic and social repression and has suffered the barbs of critics across a range of disciplines. Yet there is another dimension of the story—one in which the housing industry plays an important role in the reform saga.

One of the earliest reform strategies provides a useful illustration of the interconnectedness of the free market and the reform impulse. "Philanthropy and 5 Percent" relied on socially conscious capitalists for the philanthropy yet provided economic return to perpetuate the reform process. Each constituency saw benefit in the interaction and touted it as a model that united the spirit of uplift with the realities of the marketplace. The creation of model company towns and industrial villages also relied upon an equation of mutual benefit for capitalists and reformers. While these relationships were often neither easily categorized nor easy to duplicate, and even though some reformers were uneasy with their overtly commercial character, the movement saw potential for broader social application early on. Indeed, until the crises of war and depression impelled the state to take a more active role in housing issues, reformers had few other places to turn with such ready resources. Usefully entangling business and industry in the reform process required creative thought and careful planning but remained a potentially promising tactic in the quest to achieve productive results.

As the federal government became increasingly involved in housing issues, reformers and the housing industry faced a new situation filled with promise and fraught with potential conflicts. Many in the industry feared "creeping socialism" on a philosophical basis and the possible erosion of profits from government competition in housing construction on a more practical basis. At the same time, however, some in the business of housing saw great promise in the new era and realized that economic relationships with an activist government could potentially return handsome profits. Each sector had common interests, though identifying areas of mutual cooperation took time to develop, required creative thinking on both sides, and frequently required intermediaries who could facilitate both the process and its perceptions.

Beginning with U.S. involvement in World War I, the federal government played an increasingly significant role in expanding opportunities for home ownership. Visionary elements in industry saw a mass market for housing waiting to be conquered and were intrigued by the idea that an activist government might assist in the conquest. Even businessmen in industries not directly related to housing sought ways in which to adapt their products to housing in order to benefit from government-sponsored programs and initiatives. The often combative nature of the government/industry relationship in the housing market has masked the significant role that the housing industry has played by embracing opportunities presented by governmental policy. The form and consequence of the embrace did not satisfy the desires of any relevant constituency perfectly, but out of the interaction grew a restructuring of the American housing system that increased opportunities for home ownership in desirable community settings for greater numbers of Americans by midcentury.

Architects, businessmen, reformers, and other progressives attuned to the housing industry soon faced a circumstance ripe with possibilities. American intervention in World War I and the organization of a wartime economy brought large-scale federal involvement in housing issues for the first time. The pressing need to provide housing for defense workers in plants across the nation provided an acceptable rationale for unprecedented governmental involvement in the private housing market. Like the capitalists who planned company towns as a means to foster productivity and mitigate labor disputes, the federal government sought to achieve these goals on a nationwide level. And like the planned company towns, federal housing programs advanced a broad conception of housing and community that reflected progressive ideology.[9]

The Wilson administration initiated proactive and coordinated programs to address the housing problem as a key component in its war mobilization plan. The Department of Labor's U.S. Housing Corporation (USHC) and the U.S. Shipping Board's Emergency Fleet Corporation (EFC) were created in 1917 to organize housing policies and manage their relationship to overall labor strategy.[10] The architects, planners, and policy makers who designed the wartime communities had a clear vision of what form their projects should take. They did not replicate the urban flat or the log cabin, but the middle-class suburban community. They emulated the middle-class suburb because they believed that it held the key for advancing the social identity of the working classes. By practicing community building on a national scale, they hoped to encourage stability and patriotism as well as productivity. Policy makers promoted and justified federal intervention on grounds of increased productivity, but they also seized the opportunity to provide the working classes with the best implementation of housing and community then available. By molding the built environment of workers, the federal housing programs attempted to impose a vision of an idealized middle-class lifestyle upon the 250,000 workers and their families who occupied government-sponsored housing during the war years.[11]

World War I defined an important role for the federal government in housing reform. The connections between worker housing and productivity, social harmony, urban congestion, and the health and morals of the working classes presented the federal government with an unprecedented opportunity to influence society and impose a coherent vision for future reform. In the short period between spring 1918 and the end of 1919 the USHC and the EFC built over 16,000 houses at 150 sites throughout the country, including many complete communities. Inspired by the English Garden City movement, the USHC's Architectural Division advanced a vision of the ideal worker community that both reflected past reform efforts and influenced future thinking. They built permanent houses—over 90 percent of them were detached single-family or two-family structures—and situated them within well-planned communities on the urban periphery. Certainly economics played a primary role in this strategy, but the program also believed that permanent structures provided workers with something worth owning and reflected the best aspects of the communitarian ideal.[12]

The war housing program was both a continuation of the reformist impulse and a stunning departure from the past in terms of federal involvement. The USHC and the EFC built more than houses—they provided a model for community building as well as a blueprint for future efforts. Yet the locus of action within the federal government was problematic from the beginning and increasingly unpalatable to a nation steeped in traditional views about housing. What was possible in an atmosphere of crisis and what was possible in an atmosphere of peace and prosperity proved vastly different. Almost immediately after the armistice, Congress moved to end the programs, divested its ownership in remaining housing stock, and launched an investigation into the "egghead" influences in the USHC as part of a general backlash against governmental involvement in the private housing market.[13]

Yet the retreat by the federal government from direct intervention in the housing market did not mean that it exerted no influence over the process of residential decentralization that gained momentum throughout the 1920s. Although the tactics of direct intervention were not politically palatable outside a crisis setting, the state continued to influence the progress and nature of the American housing system. The most enduring contribution was the articulation of an ideal for American housing: the single-family house situated in a zoned, planned suburban community built by private industry. The federal government's proper role, best articulated by Herbert Hoover, was to promote the suburban ideal and facilitate the flow of capital to private industry for its physical implementation. Hoover's "associationalist state" assisted the rise of suburbia and promoted it as the locus of the American dream through the public encouragement of private initiative. Interestingly, the abandonment of direct market intervention did not include an abandonment of the wartime housing program's embrace of the suburban form as the preferred instrument for social progress through better housing.[14]

The housing industry, or "real estate lobby" as it was often described, consisting of investors, builders, developers, real estate firms, and manufacturers of housing components, enthusiastically welcomed federal endorsement of the suburban form but resisted direct involvement at every turn. The public largely supported this viewpoint, especially as fears of socialism spread in the wake of the Bolshevik revolution. In that atmosphere the prospect of a federal government deeply involved in the housing market seemed unthinkable. Indeed, the suburban homestead provided a powerful anticommunist image as the bastion of American citizenship, economic free will, and private enterprise.

Although associationalists like Hoover rejected direct intervention, they conceived a vigorous federal role as booster and facilitator of relationships between government, private industry, professional groups, and reform organizations in the interest of promoting the suburban single-family dwelling. A number of organizations and promotional efforts received federal endorsement, including Better Homes for America, the Architects' Small House Service Bureau, the Home Modernizing Bureau, and various "Own Your Own Home" campaigns. As Secretary of Commerce, Hoover encouraged these organizations and even served as president of Better Homes for America. He viewed this organization as an ideal example of the cooperative approach and regarded it as a "collateral arm" of the government in publicizing the need for better housing and home ownership, improving housing standards, and encouraging public participation in housing reform.[15] Under Hoover, the Department of Commerce also served as a research base for data collection on housing issues, sponsored housing conferences that attracted considerable industry participation, distributed literature promoting city planning, and published educational pamphlets for potential homeowners. The rationalizing impulse was also evident in efforts to promote easier access to home-building capital, review of zoning- and building-code formulation, promotion of standardized contractor specifications, and support for greater efficiency and streamlining in the lumber industry.[16]

Indeed, federal promotion of the suburban ideal provided opportunities for many constituencies. Business and industry profited handsomely from the construction of suburban communities. Home-building interests such as real estate agents, suburban developers, and contractors, as well as all their economic constituents, stood to gain dramatically. Companies manufacturing housing components benefited from increased business, as well as from the impetus for rationalization and standardization of production that inevitably accompanied an expansion of the market. As the production of each component of housing became more efficient and modern, it gave additional credence to the idea of rationalizing the entire process.

Housing-related professions also anticipated wonderful opportunities. Architects sought a greater role in directing the physical form of American housing, as did associated professions such as planners and landscape designers. Founded in 1920, the Architects' Small House Service Bureau focused on invigorating the authority of the profession in the design and placement of single-family dwellings. The Bureau naturally advocated the hiring of professional architects rather than relying on free plans from local lumber dealers or the one-size-fits-all approach of the precut house then widely available through mail-order catalogs. The infusion of "expert" talent represented an opportunity to increase the efficiency of house construction and lower costs, a benefit for the potential homeowner and a boon to the architectural profession and the building industry.[17]

The architectural profession was especially interested in demonstrating its mastery of the modern industrial process while at the same time showing what it had learned from its progressive-era experiences with community building. The profession might be just the intermediary industry that government needed. Versed in both the economic and social background of housing, architects saw themselves as uniquely positioned to carry the progressive reform vision and unite it with practical economic and political realities. The profession seemed poised to offer a heroic response to the housing problem, for it was clear by the early decades of the twentieth century that the timeworn process of home design and construction was woefully inadequate to provide affordable housing to an expanding population. The new era challenged traditional conceptions of "dwellings," demanded a thorough understanding of industrialization and system-building technologies, pushed market considerations deeper into the design process, and propelled a rendezvous with technology and social forces that promised to revitalize the architectural profession. Providing a blueprint for affordable, durable, attractive, mass-produced housing, along with the challenges of directing and influencing engineers, manufacturers, developers, and government policy makers, meant a new primacy and gravity for the profession at the center of a society with rational industrial systems in place for nearly everything except affordable housing.[18]

The influence of the associationalist approach in the furtherance of the suburban ideal provided a model for public and private sector relationships. Among these intersections lay a dynamic arena where elements of government policy, commercial enterprise, professional interests, and nonprofit organizations would vie for influence, profit, and power over the course and direction of the American housing system. In the 1920s one commonality was established: the identification of the single-family suburban home as the center of a new productive and consumptive American lifestyle. This setting would emerge to provide philosophic, political, economic, and social foundations for American domestic aspirations in the twentieth century. The promotion of the suburban homestead as the new "palace of consumption" where amenities and technical wonders opened up new possibilities for the good life influenced the course of American housing reform in meaningful ways.[19]

Whether the associationalist model could actually produce an expansion of suburban housing at a level of affordability within reach of the middle-class mass market remained to be seen. The good life of suburbia represented a worthy and desirable aspiration for individual Americas, received the blessings of the federal government, fit at least part of the communitarian vision of the progressive reformers, and presented a profitable development for the housing industry. The question of *where* seemed settled, yet the question of *how* remained a vexing prospect.

Technological Challenges

The overall expansion of the consumer economy in the 1920s brought a host of mass-produced material goods into the American household as well as an expectation for a more prosperous future. The advancement of the middle class was underway, and the spread of more sophisticated organizational and productive systems provided American industry with the means to tap this ever expanding market. The automotive industry best exemplified the transformative potential of mass production and mass marketing. By the end of the decade one in five Americans owned a car and millions more had tasted the freedom that automobility offered. However, a good many shiny new cars were parked in front of substandard dwellings—a fact that deeply disturbed observers of the state of the American housing system.

Why had American entrepreneurial vigor made the automobile affordable to the mass market but not housing? Why was housing the "industry that capitalism forgot?"[20] The appeal was seductive and provided one of the most prominent constructs for the critique of the housing industry for the next three decades. The public understood the analogy, the press seized upon its apparent simplicity, and excitement surrounding a technological solution to America's housing problems grew consistently more intense in both public and private sectors. America's housing problem, which had been a constant since the industrial revolution, seemed ready to surrender to the application of systematic thinking, modern industrial methods, and innovative architectural design. The transformational potential of affordable, mass-produced housing had broad social, political, and economic ramifications. The expansion of home ownership could provide an engine for economic expansion as well as salve a variety of social ills spawned by industrialization. Reformers had more than made a case for decent housing as a bellwether of a modern progressive society. Now it was time to employ the best means that modern society could muster to make the object of desire created by both industry and government a reality.

As the prosperity of the 1920s lifted more Americans into the middle classes and a middle-class housing market began to emerge, observers of the American housing system wrestled with a strategic shift. Increasingly, the dialogue moved from discussions about housing *construction* to a systematic analysis of housing *production*. Aligning the housing process with the promise of the machine age drew the interest of a diverse range of thinkers and doers, including prominent representatives from political, academic, professional, and commercial life. Part of a larger consideration of social organization that emerged from the intellectual ruins of the Great War, housing became a key element in the national discourse about economic progress, social stability, and political ideology.

The discussion encompassed a wide range of topics, including commercial prospects and market consequences, aesthetic appraisals, political assessments, intellectual critiques, and social considerations implicit in restructuring the productive organization of the housing industry. These "modern" thinkers were consistently troubled by the apparent unwillingness of the industry to embrace the potential of the factory age and face up to the social promise inherent in a new and powerful productive paradigm. This point of view provided a common motif for the diverse range of voices calling for an industrial solution to America's housing woes.

Business leaders were often particularly harsh in their assessment of their fellow capitalists in the housing industry. Edward Filene, a noted Boston merchant, social critic, and influential voice for progressive business practice, publicly bemoaned the overly complex and inefficient architectural design of many American houses. Custom building to individual tastes presented a major roadblock to mass production, Filene argued. "We are paying a high price for complexity," he wrote, arguing for a rational approach to housing standardization that would "eliminate everything that does not add to its beauty, its quality, or its usefulness." Excessive ornamentation and adornments emulative of older European styles struck Filene as especially useless and ill suited to the new technological promise of modern American life.[21]

Similarly, social critic and historian Lewis Mumford argued for a new American interpretation of housing that embraced a modern aesthetic while retaining meaningful connections to people's lives, their communities, and the environment. Reproducing the French chateau, the English manor house, or the Mediterranean villa not only interfered with a purely American interpretation of housing but also mitigated the efficient application of modern production methods. A new design aesthetic grounded in the productive potential of the factory system and based on a greater harmony between American lifestyles and American physical environments seemed to hold transformational promise.[22]

The suburban restructuring of residential life represented a rational and holistic view of housing and social reform for the middle classes and a hopeful vision for those aspiring to middle-class lifestyles. The integration of land, people, and industry, viewed as mutually interacting elements of social policy, provided a blueprint for the reconstruction of American society based on an ideal balance between urban and rural. Suburbia would become the preferred locus for a modern industrial and consumptive society, free from the deleterious living patterns created by the first industrial revolution and free to accept the architectural, social, and political imprint of a purely American vision for modern living. Suburban life would produce a new American—propertied, invested in the social order, and prepared to enjoy the benefits of decent housing and stable communities without abandoning the cultural and democratic advantages of the urban experience. Indeed, Mumford's views represented the belief that the community development model of progressive housing reform could be brought to full fruition only in a suburban setting.[23]

A grand vision indeed, but who would build this new utopia? And would anyone of modest means be able to afford it? The political counterreaction to the federal government's direct involvement in the housing market during World War I precluded a prominent governmental role in the creation of the new suburbia. Real estate and construction interests stood fast to the view that government should support but not compete with the private housing market. Yet experimentation with hybrid approaches to model suburban community building revealed a willingness on both sides to pursue innovative ideas. Model suburban communities presented reform organizations, commercial interests, and governmental entities with opportunities to transform vision into reality. Using limited dividend strategies, working in concert with local and regional government, and attracting financial support from progressive business interests, model communities such as Chatham Village in Pittsburgh, Pennsylvania, Sunnyside Gardens in Queens, New York, and Radburn, New Jersey, offered attractive, affordable, and community-rich habitats for their middle-income occupants.[24]

As desirable as these communities were, they remained "demonstration samples," limited in scope and based on financial models that were unrealistic for society at large. Mumford astutely observed that comprehensive housing reform required a fundamental transformation of the economic and political underpinnings of the American housing system. However, changing long traditions of free enterprise and private land speculation, as well as restructuring all of the institutions that supported them, seemed a near impossible task. "Modern housing," Mumford wrote, "demands not merely an improvement of the physical structures and the communal patterns. It demands such social and economic changes as will make it available to every income group."[25] Reformers like Mumford and Catherine Bauer, a longtime housing activist and author, believed that a suitable remedy to the "boutique" nature of the housing industry lay in the realm of the federal government. Bauer observed that there

> must come a technique for building complete communities designed and administered as functional and constructed by large scale methods. Only government can

> make the decisive steps and set up a new method of house production as a long time social investment to replace the wasteful and obsolete chaos still prevailing.[26]

The housing industry had not matured in an industrial or societal sense. It had not produced a product suited to the needs of the mass market. It had not followed the Ford model of industrial progress. In short, capitalism had failed the housing industry, and the housing industry had failed capitalism.

What was the private housing market doing to develop the productive technologies necessary to build a suburban experience that would be accessible to the mass market? Although the industry had made important technological strides since the turn of the century, it was still far from offering an elegant, affordable, and practical solution to the integration of productive technologies and community building. However, innovative firms and individuals were working toward better manufacturing processes and wrestling with a more systematic view of housing production. Consideration of alternative building materials such as concrete and steel, as well as experimentation with composite wood products, grew more prevalent and attracted considerable attention from reformers and entrepreneurs.

For example, Thomas Edison proposed a system for constructing an entire concrete house in one pour operation in the first decade of the century. While the "monolithic concrete house" proved impractical on a commercial scale, it reflected increasing interest in prefabricated construction techniques using industrially pliant materials. Architect Grosvenor Atterbury, another early pioneer, devised a housing system of precast hollow-core concrete panels for floors, walls, and roofs. Supported by the Russell Sage Foundation, Atterbury erected several hundred houses in Forest Hills, New York, between 1910 and 1918. Although the process also proved too expensive for widespread commercial implementation, Atterbury's work was significant because it was the first philanthropically supported experimentation in prefabricated construction for the mass market. The Russell Sage Foundation made the results of the Forest Hills project, including patents for specific construction innovations, available to other nonprofit organizations working toward "a scientific solution of the housing problem."[27]

The idea that prefabricated construction methods using alternative materials might provide the solution to the housing needs of the majority of Americans represented a change from previous views of prefabrication as suitable only for temporary construction. Prefabricated dwellings had a long history of filling housing gaps in newly settled areas or as secondary shelter for recreational purposes. The challenge before industry pioneers like Atterbury was how best to move prefabrication from a specialized temporary housing solution to a "movement" that would influence the progress of industrialization in the housing industry and encourage a more systematic view of the American housing system. Reconciling technological progress with reformist thinking about housing and community also represented a difficult challenge.

In the coming decades, prefabrication would gain credibility, though progress was slow due to several factors. The decentralized structure of the housing industry made it difficult for individual firms to undertake research and development efforts into new construction systems. Capital-starved and existing from project to project, few builders had the resources or perceived the need to experiment with alternative materials or processes before they were proven cost effective and had gained widespread public acceptance. Larger corporations in the building products industry also saw few benefits in developing overall construction systems unless those systems used their products exclusively, which was unlikely. Although firms continued to refine the production of building components, the challenge of developing industrialized building systems far exceeded the scope of individual organizations. The overall health of the housing market until the mid-1920s also mitigated development since companies had few incentives to pursue new methods in the midst of unprecedented prosperity.

Perhaps most important, however, was the lack of governmental stimulus and technical assistance to the housing industry. Until the 1930s the federal government did virtually nothing except for a small program in the Bureau of Standards that encouraged the standardization and simplification of building materials. In contrast, European governments actively pursued centralized policies for public housing and sponsored research and development projects to spur the acceptance and marketability of new building systems and construction materials.[28] As we have seen, this was certainly not the American way, and neither the federal government nor the housing industry had any systematic interest in industrialized building until the advent of economic crisis. Consequently, pioneering individuals and enterprising firms pursued research and experimentation in industrialized building largely on their own.

One industrialist who heeded Edward Filene's and Lewis Mumford's calls for technological advancement was Albert Farwell Bemis. As head of Boston-based Bemis Industries, a firm that controlled various building-product concerns, Bemis funded research into prefabricated construction systems using materials such as plywood, concrete, and steel. Throughout the 1920s his firm experimented with various systems and amassed a large amount of data that influenced future development. Although no single system proved commercially viable, research into elements of the building process refined thinking about prefabricated design. Specifically, Bemis advanced a "cubical modular" design theory that stressed flexible coordination of structural components. Work on joining methods proved especially promising, since a reliable system for tying construction modules together represented the most challenging aspect of production because it had tremendous impact on labor costs regardless of the type of materials employed.[29]

Bemis published the results of his company's work in a three-volume series entitled *The Evolving House*. The first two volumes surveyed the history and economics of housing. The third volume, *Rational Design,* provided a theoretical discussion of modular design theory and offered practical guidance on the dimensional engineering of modules and advice for architecting a "continuum of cubes" to achieve maximum cost effectiveness and design appeal. Bemis believed that the use of "stock" modules coupled with creative architecture provided the key to developing systems that improved quality, reduced cost, and lent themselves to mass market production.[30]

Eager to demonstrate production economies inherent in modular building components, Bemis Industries offered technical assistance to firms willing to put its research into the field. Several companies sought partnerships with Bemis and built small-scale developments using the approach, but none attempted a large development since financial institutions were unwilling to support such "radical" and unproven construction methods. Bemis's work had a greater long-range impact on the housing industry through the company's creation of the Modular Service Association (MSA), a nonprofit organization dedicated to the promotion of dimensional coordination and standardization of construction materials and processes. Industry groups such as the American Standards Association and the American Institute of Architects served as joint sponsors of the MSA. In the 1930s and 1940s, the MSA worked with the building industry and the federal government to advance "modular coordination" and to promote the reduction of labor-intensive and costly on-site fabrication. Bemis was indeed a visionary who legitimized basic concepts of prefabrication within the housing industry and gave physical form to construction systems that held commercial promise.[31]

While Bemis approached industrialization within accepted architectural and construction paradigms, other pioneers advanced more revolutionary views of housing and technology. Buckminster Fuller's Dymaxion house of the late 1920s represented a reconceptualization of housing and the processes that produce it. Fuller argued that housing design had to conform to modern industrial methods rather than attempt to recast modern industrial methods to accommodate outmoded notions of housing forms. The original Dymaxion house was a prefabricated hexagonal metal unit suspended by wires from a central mast and included numerous amenities. Decades ahead of available production technologies, the house never progressed beyond the model stage. However, its radical nature excited the public about the promise of a new era in housing based on technological progress and challenged architects and engineers to consider how housing design could be reconciled with emerging building technologies.

The Dymaxion would be the first of many "houses of the future" to capture the public imagination and spur ongoing development of innovative housing technologies.[32]

Although elements of the housing industry began to conceptualize a role in the reconstruction of the housing system, uniting reformist visions with practical implementation demanded radical changes that industry alone could not achieve. It would take a crisis to impel action and an activist federal government to guide the process. The tumult of the Great Depression and World War II focused American institutions on the need to reconstruct the housing industry, expand the role of the federal government in guaranteeing opportunities for decent housing across the income spectrum, and implement policies to expand home ownership among the middle classes. Frank Lloyd Wright observed that "a citizen in his own life in his own home with his feet on the ground . . . is truly a free man."[33] The connection between decent housing and "freedom" in America would take on a new meaning in the years to come as the nation faced a series of enemies that challenged the definition of American ideals. If home and hearth were to feed the nation's strength as it battled the evils of depression, Nazism, Japanese militarism, and communism, home and hearth needed far more attention.

Housing the Commonweal

As the economy disintegrated after the Great Crash and the housing market collapsed, the nation found itself woefully unprepared to address a housing crisis fueled by a growing number of foreclosures. Failing credit markets took a heavy toll on those families who were able to purchase homes for the first time during the 1920s. The promise of middle-class status through suburban home ownership so deliberately trumpeted by the government and the housing industry gave way to the grim reality of downward mobility. Foreclosures threatened to destroy the image of a prosperous middle class as a key to national stability and progress.[34]

The federal government was initially unable to respond the crisis. It had no strategic approach to housing policy, no programs in place to assist homeowners who faced foreclosure, and no programs to aid the perpetual problem of access to decent shelter for lower income families. The freewheeling economic ideology of the 1920s and the success of the housing industry in forestalling governmental "interference" in the housing market now seemed absurd in the face of a crisis that reached deep into the heart of the American middle class. A new political climate that expected—even demanded—federal action brought new pressures and opportunities for the housing industry. Fruitful collaborations would emerge during the 1930s and advance housing technologies. Government, academic, and private interests devoted considerable energy to the notion that the key to America's housing problems lay in the refinement and application of productive technologies.

Issues of housing reform attained prominence during the Great Depression because they affected more Americans with political influence and the will to use it. It was not just the tenement dwellers and their supporters battling for reform. Now middle- and working-class families whose faith in the American way had been shaken demanded action and petitioned government for reform and redress. For the first time in American history, public faith in the housing industry began to crumble, and the industry prepared for an unprecedented assault as people turned to the federal government for strategic, broad-based solutions.

Reformers rejected the claims of the real estate lobby that the housing crisis was solely the result of weakness in the housing market since the mid-1920s. The reform community argued that the problem went far deeper and was rooted in the structure of the industry. Inefficient and often low-quality production practices implemented by small firms using multiple subcontractors with minimal oversight, coupled with unnecessarily complex financing and valuation practices, defined a system hopelessly mired in the preindustrial past and unable to provide suitable housing even in good economic times. In contrast, real estate interests saw strengths where reformers saw weakness. The complex web of economic relationships in the housing industry reflected the true spirit of American free enterprise and one of the last bastions of individual entrepreneurship in an economy increasingly dominated by large-scale enterprise. Like the mom-and-pop retail establishments that resisted the advent of large chain stores in the 1920s and 1930s by lobbying government for protective legislation, the real estate community felt that government should protect the status quo of the housing industry since it virtually defined all that was good about the American system.[35]

Yet the economic and social pressures of the depression and the centrality of the housing industry to economic recovery made such views untenable in the public mind and among policy makers in the Roosevelt administration. Roosevelt's Brain Trust turned first to strategies designed to stabilize the housing market by restructuring the nation's home finance credit system. This was politically efficacious since easing foreclosures meant tying an important constituency to the New Deal in its early stages. The administration would also pursue housing policies designed to assist a greater spectrum of Americans by expanding opportunities for home ownership through government-sponsored community development, some of which produced beautifully designed and socially vibrant communities.[36]

Less visible to the public but prominent across a range of initiatives was support for the development of housing technologies. Veteran reformers who argued for technological progress during the 1920s assumed important positions in New Deal housing agencies and consistently advocated the application of mass-production methods to the housing market. Improving building technologies served multiple purposes: as a means to improve the industry's capacity to serve the mass market; as an incentive to create jobs in an industry fundamental to economic recovery; as a strategy to effect change in a manner more palatable to the industry than broad regulation or outright competition; and as an example of government-supported efforts that assisted capital-starved firms in the research and development of new technologies and products that held future promise.

Indeed, much of the federal government's housing efforts during the New Deal can be characterized as subsidies to the housing industry in one form or another. The New Deal was actually quite gentle on the industry. While controversy frequently arose over specific measures, commercial interests rarely failed to feast at the federal trough—even though they consistently denounced federal involvement as socialistic. What emerged was a spectrum of subsidy and involvement, from constructing residences, developments, and communities, to assisting technological development, implementing construction standards, and stabilizing credit markets. Governmental "control" was easily identified but not always appreciated. Using the federal system to promote the rational productive capacity of the housing industry was the greatest contribution of the New Deal to the industry and set the stage for the expansion of better housing opportunities for a greater number of Americans in the decades to come.

The New Dealers first turned to the Home Owners Loan Corporation (HOLC), an agency created by the Hoover administration in 1932. HOLC emerged from the president's Conference on Home Building and Home Ownership held in December 1931. The conference endorsed Hoover's plans for establishing a system of home loan support banks similar to the Federal Reserve System in an effort to back primary mortgage-lending firms and restore faith in the mortgage credit system. HOLC addressed the issue of home foreclosures, and the politically savvy New Dealers greatly expanded its scope. Backed by federal capital and bonds, HOLC lent money to savings and loans to purchase defaulted mortgages and refinance low-cost loans to homeowners. Thousands of Americans were able to retain ownership of their homes with help from HOLC, and many of them became lifelong Roosevelt supporters.[37]

The administration soon advanced a more ambitious program with the Housing Act of 1934, which established the Federal Housing Administration (FHA). The FHA sought to reorder the housing industry by promoting the construction and financing of moderately priced single-family houses and by implementing standards for community planning in suburban settings. The FHA insured long-term mortgage loans made by private lenders as well as loans financing home improvements and repairs. The Act also authorized the establishment of national mortgage associations to provide a secondary market for home mortgages. The FHA-insured mortgage transformed the American housing system by insuring billions of dollars of mortgages and bringing much needed stability to the housing market. It took much of the speculative risk out of the equation for builders and thus made possible the financing of large-scale developments.[38] The FHA helped the housing industry move toward rationalization of the financial aspects of the business. Rationalizing the production end of the equation, however, would take additional time and considerable effort.

HOLC and FHA demonstrated FDR's commitment to the housing needs of the middle class. But these measures did little to assist the "one-third of a nation" that remained ill-housed. New Dealers who advocated a greater role for the federal government in the financing and construction of houses for the poor and the working classes exerted substantial influence in the creation of policies that pushed the administration beyond a purely fiscal approach focused on the middle class. The economic crisis enabled reformers in and out of government to call for direct, large-scale housing initiatives for lower-income Americans. The private housing market had failed, they argued, and furthermore demonstrated little interest in providing decent low-cost housing. In their view government involvement in the low-cost market was not competition, but a necessary extension of the industry for the public good.[39]

The passage of the National Industrial Recovery Act (NIRA) moved the federal government deeper into the housing industry than ever before. The Housing Division of the Public Works Administration (PWA) built housing for low-income working-class families between 1933 and 1937. Designed as a temporary agency, it nevertheless produced well-designed and attractive developments. In collaboration with the FHA and the local housing authorities in Fort Wayne, Indiana, the PWA also explored the use of prefabricated stressed-skin wood panels to construct single-family houses. Known as the Fort Wayne Plan, the development used a combination of public effort and private financing to encourage the proliferation of affordable housing. The PWA Housing Division provided thousands of jobs and economic stimulus in fifty-nine communities across America. The quality of construction, upscale materials, and artistic touches by PWA craftsmen gave these communities a dignity previously unknown in lower-income housing.[40]

However, the PWA Housing Division was intensely controversial from the start. Led by Harold Ickes, the PWA was designed to build housing for those who could not afford privately built homes. Yet industry groups such as the National Association of Real Estate Boards, the United States Savings and Loan League, and the National Association of Retail Lumber Dealers stalwartly opposed the PWA and viewed it as a threat to the free enterprise system. In their view HOLC and the FHA were appropriate measures to stabilize the private market, but any further governmental involvement was unwarranted.[41]

"Despite industry opposition, the New Dealers pressed on. Buoyed by FDR's landslide reelection in 1936 and the continuing economic crisis, the progressive coalition pressed for the creation of a permanent governmental agency that would continue the work of the PWA. In 1935 they began to fight for the Wagner Steagall Act, which called for the creation of a new permanent housing agency. Congress passed the Act by an overwhelming margin, and in September 1937 Roosevelt signed it into law. Wagner Steagall stated that the provision of good housing for the poor was a perpetual social obligation and created the U.S. Housing Authority to implement the new social vision—"the provision of decent, safe, and sanitary dwellings for families of low incomes and the eradication of slums." Under the direction of Nathan Straus, Jr., the USHA assembled a talented and experienced staff of reformers, including Catherine Bauer and Leon Keyserling. The Authority operated by funding local housing agencies and subsidizing construction and maintenance costs. By 1939 the USHA had eclipsed the efforts of the PWA, constructing over 50,000 public housing units. In addition to the overall benefit of providing needed housing and stimulating the construction industry, the USHA's focus on housing technologies exerted an important influence on the industry at large.[42]

Additional government agencies and programs that provided leadership in the development of affordable housing using advanced methods and materials included the Resettlement Administration and its successor agency the Farm Security Administration, the Tennessee Valley Authority, the U.S. Forest Products Laboratory, and the National Bureau of Standards. These agencies created and implemented sophisticated economic models that assessed all elements of the construction process in an effort to refine and rationalize. Assessment of labor practices, raw material costs, production methods, and process efficiencies gave planners a clear picture of costs associated with each element of the project. Emphasis on prefabricated components, division of labor, and innovative material handling systems seemed finally to bring the benefits of modern management to the housing industry. The managerial approach provided a lesson to the housing industry that would have significant consequences in the coming decades.

Established in 1935, the Resettlement Administration developed housing for low-income families. Its most notable contribution was the creation of the "greenbelt" communities, which attempted to adapt the English garden city approach to the housing needs of families displaced by the depression. Located thirteen miles from the heart of the nation's capital, Greenbelt, Maryland, was the administration's first effort. This development, and the two others in Greenhills, Ohio, and Greendale, Wisconsin, became laboratories in which planners, architects, landscape contractors, and construction firms experimented with new ideas, processes, and building technologies. Prefabrication played an important role in the construction of town houses, apartment buildings, and single-family residences. Assembly line construction methods in the field and the use of alternative building materials provided excellent examples for the private building industry. Encompassing the best of the communitarian ideal, the attractive layout of the towns produced a vibrant social experience and demonstrated that innovative and cost-effective building methods could produce an attractive result.[43]

This was government involvement at its best. *Architectural Forum* reminded the housing industry that the greenbelt program offered a powerful blueprint for widespread implementation: "private builders may well study [the Resettlement Administration] for clues to . . . simplify and therefore generally improve standards of planning, methods of design, and construction."[44] Several private builders indeed drew valuable lessons from the greenbelt communities. Abraham Levitt and Sons used many of the same planning and construction techniques to create middle-income housing on Long Island in the late 1930s. Using a holistic approach that attempted to rationalize all elements of the development process, Levitt controlled everything from land acquisition to the furnishing of individual houses in an effort to provide "swank" at a reasonable cost. Similarly, in Chicago, William Harmon used the same approach to develop housing for wage-earning families. Harmon's Colonial Village was immensely popular and served as a model for the future of mass suburban development.[45]

In 1937 the Farm Security Administration (FSA) replaced the Resettlement Administration but continued its mandate. The FSA advanced construction technology by researching and employing stressed-skin plywood prefabrication in the construction of housing for displaced farm workers. By 1940 it constructed over 20,000 individual units and developed significant expertise in site fabrication, which would prove essential to the success of the defense housing program during World War II. *Architectural Forum* noted that "in the face of national emergency, Farm Security stands out as the agency most experienced in the work of building houses quickly and cheaply."[46]

The Tennessee Valley Authority (TVA) also made significant contributions to the advancement of housing technologies and construction processes during the 1930s. Housing thousands of construction workers at numerous project sites presented a challenge that had to be met quickly. In 1934 TVA architects began developing sectional housing that could be transported by overland trucking. Using wood frame construction, the houses were prefabricated to a high degree of finish in the factory, including all mechanical systems and even such minor components as light bulbs and screens. Sections were bolted together at the erection site and prepared for habitation in as little as four hours. The reduction of site labor resulting from the high level of factory fabrication saved the TVA millions of dollars and provided a lesson in the potential of full factory prefabrication that was not lost on private industry. Prefabricators looking to bring the advantages of sectional housing to the middle-class marketplace found the TVA's experience invaluable. The development of the mobile home industry and subsequent advancements in modular housing can be traced to the TVA's housing program.[47]

Two other agencies of the federal government provided essential assistance to the development of prefabricated housing technologies during the depression era. The U.S. Forest Products Laboratory of the Department of Agriculture was instrumental in advancing stressed-skin wood panel construction. Researching component materials such as glues and plywoods, assessing the performance of stressed-skin panels across a variety of environmental conditions, and investigating potential long-term problems of the system provided prospective manufacturers with essential data. The industry readily embraced this type of public subsidy, since capital-starved startup firms could not afford the type of research and development expertise offered by the Laboratory.[48]

The National Bureau of Standards also provided essential technological assistance and organization to the industry. In 1937 the Bureau began a program to assess various building materials and structural systems likely to assist the development of low-cost housing. Performance testing in the laboratory and in the field aided the development of commercial standards and "simplified practice recommendations" for prefabricated as well as traditional construction. Of particular importance was research into the performance of prefabricated components and systems, which aided the struggle to reform antiquated building codes. Building codes were most commonly based on the specification of certain materials rather than the performance of overall systems. Countless local variations presented a significant barrier to prefabrication, especially since local codes were designed to favor certain materials and construction practices. Specification codes precluded the development and implementation of alternative materials and processes deemed threatening to traditional practice. The Bureau's work lent credibility to calls for code reform from the prefabricated housing industry as it began a long struggle to challenge the economic, political, and legal protection afforded traditional residential construction.[49]

Savvy builders throughout the nation took careful note of the emphasis on production efficiency, economics of scale, and technological innovation brought to the forefront by the federal government's involvement in housing. However, what caught their attention was not the opportunity to construct private developments for the poorest third of the nation, but the prospect of applying these managerial and productive technologies to the middle-class market where profit margins would be far more attractive. Private interests now had a blueprint for large-scale suburban development, the FHA to stabilize the credit system by underwriting the speculative risk, and a powerful ideological argument in promoting home ownership.

Yet the housing industry was not publicly appreciative. It continued to perceive governmental involvement in the housing market as potential competition and further evidence of the "creeping socialism" of the New Deal. It was a virulent disease, they argued, a first step in the socialization of the country and a threat to all sectors of the industry. "We do not concede the need or the right of the federal government . . . to enter the housing field by building, owning, or operating permanent housing projects," stated Joseph Merrion of the National Association of Home Builders.[50] This stance was not new, nor would it change in the coming years.

The battle over public and private housing during the 1930s set the stage for decades of controversy. Yet it was an exciting time for housing reformers and the industry in general. An activist government, advances in production technologies, and an industry eager for recovery brought constant energy to the field. Indeed, the Progressive Era looked static compared with the vibrancy of the New Deal. Innovative public projects as well as advancements by private interests set the stage for progress unimaginable in previous eras. Industrialized building seemed to come of age just as American society stood ready to embrace new ideas and approaches to the ever present housing problem.

A Systematic State of Mind

Government and industry grew increasingly interested in prefabricated building systems as depression conditions worsened. Statistics revealed the shocking truth that only 20 percent of American families in the mid-1930s could afford even a "low-end" single-family house costing around $4,000. As the government could not possibly close the gap with public housing initiatives, facilitating the development of productive technologies and pressing industry where possible to adopt them in the field seemed a reasonable approach. Encouraging other depression-crippled industries to consider housing applications for their products also fit into the administration's recovery strategy. For example, with the steel industry operating at severely diminished capacity, developing housing applications for steel products, such as framing and roofing systems, or pursuing the development of all-steel housing, presented intriguing market possibilities.[51]

The New Deal intended to "jump start" industry and expand opportunities for greater numbers of Americans to become active consumers. "Get Building Going" became a rallying cry of the administration, and industrialized building seemed an attractive element of the greater pump-priming efforts of New Deal economic policy. The challenge to the housing industry was clear: could it live up to expectations for more affordable housing through greater production efficiency? Expectations nurtured by housing reformers—industry pioneers like Bemis, visionaries like Fuller, and social critics like Mumford—coupled with an activist government focused on housing issues, now assumed a more concrete form in the public mind.

During the 1930s a number of technologies matured sufficiently to merit serious consideration for use in industrialized building systems. Lumber manufacturers developed better forms of plywood and experimented with "sandwich" systems known as stressed-skin panels, which combined framing, interior and exterior walls, and insulation into one unit. Closely related was the development of wallboard, which was intended to replace labor-intensive lath and plaster construction. Wallboards made from such materials as wood pulp, gypsum, newsprint, plant fibers, and asbestos promised to revolutionize interior finishing and also to serve as important components in sandwich construction modules. Prefabrication using wood and wood-related products garnered much attention from architects and building firms since these materials were familiar to the trade as well as amenable to their toolboxes.[52]

Other materials such as steel and concrete, though more radical in nature, also gained widespread attention due to their inherent flexibility and durability. Better quality steel and the availability of sheet steel opened up new opportunities for specialized fabrication. Improved alloys also produced lighter steel without reducing tensile strength, which would become an important factor in introducing load-bearing steel wall units in prefabricated designs. Residential construction with concrete had also benefited from technological advances. Improved mixes, especially the use of lightweight aggregates, and advanced processes for pouring and forming, gave greater dimension to concrete as a material suitable for mass-produced housing. Although both materials had significant limitations, development and experimentation continued due to the productive flexibility and potential economies inherent in large-scale application.[53]

The architectural profession watched the development of alternative housing products and construction processes closely and played an important role in mediating new technologies with traditional approaches to housing and the influences of communitarian thinking. Interest in prefabrication was widespread in the architectural profession, and it embraced the economic and social challenges of housing technologies with great vigor and much insight. Many prominent architects believed that without the guiding hand of the profession, industry would seek to impose systems of prefabrication that emphasized production efficiency at the expense of design flexibility and environmental and community contexts. Yet they remained fascinated by the potential benefits of industrialized housing and sought to exert influence over the conception and implementation of housing technologies in public and private spheres.

Walter Gropius, a German architect of Bauhaus fame who immigrated to the United States in 1937, reflected upon the tension between production efficiency through standardization and the individualistic character of American housing. Gropius embraced standardization, but warned of its impact on self-expression. He advocated a flexible system of prefabrication that would capitalize on industrial processes yet satisfy the vagaries of personal taste and fluctuating fashions of the American consumer. Like his colleague Le Corbusier, Gropius saw the house as a "machine for living" and sought design systems that achieved "uniformity in detail and variety in the general effect." Like Bemis and his "cubical modular" system, Gropius wrestled with the question of how best to design differential styles with standardized components. The art of the architect and the science of production would intersect to produce industrialized housing that was not only affordable but contextually appropriate.[54]

"Prefabrication needs the architect," implored architect Miles Colean, and indeed the best and the brightest of the profession such as Frank Lloyd Wright, Richard Neutra, George Frederick Keck, Carl Koch, and a host of others advanced prefabricated designs and significantly influenced the progress of the "movement" in the coming decades. The influence of a new generation of architects and housing engineers who attained professional maturity in the depression era also spurred interest in prefabrication and brought the enthusiasm and energy of youthful ambition to the field. Inspired by a creative freedom implicit in the design process, anxious to tackle large-scale problems, and convinced of the primacy of the architectural profession in the struggle to improve American housing, these individuals and firms explored new territory on all levels of the design process, from overall purpose to the use of nontraditional materials to physical setting. The involvement of noted architects also garnered considerable media attention and sparked the public mind with a long procession of "houses of the future." While this proved a double-edged sword in heightening public expectations, it also fueled a more sophisticated and realistic assessment of the promise and limits of housing technologies.[55]

In 1935 the editor of *House and Garden* magazine sponsored a Symposium on Prefabrication. One of the first venues to provide a forum for discussion of the myriad aspects of industrialized housing, it drew participants from various fields, including architects, engineers, businessmen, and representatives from philanthropic organizations involved in housing technologies. Conferences and round tables of various types appeared more frequently in subsequent years and served as important places for debate, discussion, and cross-pollination of ideas and people from government and industry.[56]

Conferences sparked media interest in prefabrication as well, and the majority of articles cast a positive and hopeful tone. Seemingly any new idea about mass-produced housing received serious consideration regardless of its technical or commercial feasibility. Feeding the nation's desire for solutions to the housing crisis, the media supported prefabrication but often misled the public about the immediacy and cost impact of specific designs. As one industry observer said, the press "produced words much faster than prefabrication did houses." Overall, though, media attention was extremely beneficial to the progress of prefabrication.[57]

The scope of serious efforts to advance housing technologies encompassed a wide range of public and private institutions and entities during the 1930s. Nonprofit foundations, universities, government agencies, and commercial firms large and small made significant contributions to the progress of industrialized building. For example, the Housing Research Division of the John B. Pierce Foundation continued the contributions of the nonprofit sector to housing research. Building upon the work of the Bemis Foundation and the Russell Sage Foundation, the Pierce Foundation focused on cost-efficient materials and structural systems. Founded in 1924 to foster research and development in heating, ventilation, and sanitary systems, as well as to educate the public and chronicle industry progress, the Pierce Foundation investigated a wide range of processes and materials. By the early 1930s its efforts focused on construction materials that could provide enclosure as well as structural integrity. Using a combination of steel framing and a hydro-calcium silicate composite known as Microporite, by mid-decade the Foundation had constructed several promising prototypes.[58]

Partnerships between universities and building firms also aided the development of prefabricated technologies. Purdue University founded the Housing Research Project in 1935 to investigate new designs for low-cost housing. Detailed engineering studies of overall designs and various subcomponent systems provided industry with valuable data and influenced future designs. Universities and foundations also played important roles in the consideration of housing issues beyond physical structure. Within the context of serious research, these institutions brought a value-added proposition to the movement that industry found difficult to accomplish. The fragmentation of the housing industry and the many disciplines involved in construction processes often prevented due consideration of factors external to production. With commercial pressures removed, a more holistic approach to prefabrication could emerge.[59]

The commercial promise of mass-produced housing drove both entrepreneurs and established firms toward more efficient systems during the 1930s. Investment increased as the media lavished attention on new innovations and industry personalities. By the end of the decade, approximately forty U.S. firms manufactured prefabricated houses. However, they contributed less than 1 percent of the single-family houses built between 1935 and 1940.[60] The industry was still immature, highly speculative, and, in terms of materials used, somewhat conservative. The industry also focused far more on production than distribution and marketing, thus leaving the question of how best to integrate all facets of industrialized building for future endeavors.

Because the development and implementation of prefabricated housing was capital intensive, it is not surprising that large firms took important steps toward commercialization. Interested in the forward integration of current product lines, large firms hoped that mass-produced housing could create a new market and fuel corporate recovery. They experimented with housing on various levels, from developing prefabricated mechanical systems to designing entire houses and communities. Corporations such as U.S. Steel, Johns-Manville, Pullman Standard, Celotex, U.S. Gypsum, American Rolling Mill (ARMCO), Bethlehem Steel, and General Electric pursued interest in prefabrication. Big business advanced the industry in several important ways. The advent of managerial specialization and organizational theory then blossoming throughout American industry provided important tools to define the scope and scale of production and distribution problems facing the nascent prefabricated housing industry. Large firms provided the capital infusion and human resources necessary to nurture technological advancement. Engineers were particularly important for the advancement of production standards and quality control. Integrative thinking and strategic analysis were difficult concepts for the housing industry. Prefabricated housing required the mastery of industrial systems and a keen understanding of market behavior to achieve success. The participation of successful large firms helped the industry attain a greater understanding of its potential and a clearer picture of the obstacles inhibiting its further development.[61]

However, big business did not have the solution to the transformation of the American housing system. Industrializing the housing industry was a process far exceeding the capacity of any single institution. Yet several large-scale ventures during the 1930s demonstrated the influences of systematic thinking and foreshadowed future trends. General Houses, Inc., organized in 1932 by Howard T. Fisher, advanced the idea of using prefabricated standardized components for constructing custom-built houses. The significance of General Houses was its hybrid approach. Modeled after the automobile industry, it organized a system of prefabrication that employed external vendors to supply standardized components, which it then assembled under the General Houses brand name. Unlike the automobile industry, however, General Houses had no factory. Instead, suppliers shipped parts to regional warehouses for coordination, and the company then shipped the component packages to construction sites.[62]

General Houses employed a steel framing system with sandwich panels of asbestos, cement, insulation, and plywood for the interior walls. Prices for the house in 1934, not including freight and the cost of a lot, ranged from $4,500 for a four-room design to $8,550 for a six-room, two-story model with attached garage. Suppliers included Bethlehem Steel, Weyerhaeuser Timber, and Pullman Standard. Although General Houses largely avoided technical problems in the manufacturing and distribution process, financial and marketing difficulties prevented the firm from achieving the full benefits of its business strategy. Foreshadowing key barriers for the industry, General Houses faced opposition from the building trades, where the use of prefabricated components was perceived as a threat to traditional craftsmanship, and from financial institutions, which had no valuation models for prefabricated housing and thus shied away from writing mortgages.[63] Although it achieved limited success in the private market, General Houses became one of the first prefabricated housing firms to participate in the defense housing program where risks of unfavorable labor relations and unpredictable financial valuations were assumed by the government.

Another venture that influenced the development of the industry was Houses, Inc., a corporation founded by construction engineer Foster Gunnison in 1934 at the request of Owen D. Young, chairman of General Electric. Gunnison and Young did not intend to produce houses but envisioned the company as a clearing house, liaison, and facilitator for the development of new prefabricated systems. With financial backing from a number of building-product firms, Houses, Inc. proposed to conduct research, raise capital, and provide technical and managerial assistance. One of the most notable ventures of Houses, Inc. was the "American Motohome," designed by American Houses, Inc. The Motohome, a steel frame and asbestos panel house, incorporated mechanical systems developed from Pierce Foundation research and innovations at General Electric. The debut of the Motohome in April 1935 reflected the keen interest of the press and the public in prefabricated housing and established a pattern for product introductions followed by subsequent firms. The "unveiling" of the Motohome took place in Wanamaker's department store in New York City, where President Roosevelt's mother, Sara, officially opened the exhibition. During the next six months nearly 500,000 peopled toured "the prefabricated house that comes complete with food in the kitchen." Model homes erected in nearby Westchester County also drew extensive press coverage and thousands of visitors. Company advertising illustrated "Motohome Communities" and depicted fleets of trucks carrying this technological wonder from factory to home site.[64]

Houses, Inc. invested in other designs, but disputes among company management and strained investor relations eventually doomed the enterprise. Gunnison left the company in 1935 to start his own prefabricated housing firm, and a year later General Electric abandoned its interest, forcing Houses, Inc. to declare bankruptcy. Yet two firms associated with Houses, Inc. continued to pioneer prefabricated designs and refine business models. Although American Houses abandoned production of the Motohome due to low sales, it continued to develop prefabricated technologies for the low-cost market. Focusing on wood rather than steel, American Houses was able to pursue partnerships with project builders in an effort to build sales volume. The move to wood prefabrication made its product more palatable to consumers and required less sophisticated production equipment. Better able to rationalize its productive investment, American Houses became an important contributor to the defense housing program during World War II.[65]

Additionally, Foster Gunnison parlayed his experiences with Houses, Inc. into one of the most important prefabricated housing firms in the nation. Gunnison Homes, Inc. built stressed-skin plywood panel homes for the low-cost market and achieved notoriety as an innovator in production methods and marketing techniques. A master promoter and engaging personality, Gunnison understood the marketing challenges faced by prefabricated housing firms and devised ideas to encourage acceptance of his product. One idea, the "Gunnison Village Plan," used FHA-insured mortgages to encourage local project builders to acquire land, erect a community of Gunnison houses, and provide amenities such as parks, playgrounds, and community centers. His attempt to integrate prefabrication with conventional merchant building proved too risky for prospective builders during the depression, so the company developed a marketing and distribution system using retail dealers selling to individual customers. The latter system spread the risk of acceptance over a wider range of potential customers, many of whom might like the house but hesitate to live in a "prefabricated community."[66]

Gunnison's focus on marketing represented an important step for the industry, and the Gunnison Village Plan represented a keen understanding of the problems of integrating technologically advanced production with established systems for residential development. Although production innovation was important, prefabricators who ignored the realities of the real estate industry were likely to fail. They might have the best house in the nation, but they would have nowhere to put it. As Gunnison's case illustrates, the franchise dealer system served as an interim solution to the marketing and distribution dilemma. Influenced by the effectiveness of the dealer system in the automobile industry, prefabricators believed that such a system would carry them through the initial phase of development until widespread acceptance in the financial industry would pave the way for large-scale development projects and partnerships with merchant builders.

While the heirs of Houses, Inc. found success using wood, other firms remained convinced that steel offered the most flexible and potentially efficient material for prefabricated housing. In 1932 the Ferro Enamel Corporation and ARMCO Steel introduced the all-steel ARMCO-Ferro House. A frameless design, the house employed a chassis system of large porcelain-enameled steel panels that bore the weight of the structure. Enameled-steel roofing shingles complemented the wall panels. Even interior features such as baseboards, lighting fixtures, switch plates, and flooring were fabricated from porcelain-enameled steel. Built to demonstrate the potential of steel housing, the press proclaimed the ARMCO-Ferro house the "highlight of the exhibit of all 'homes of the future'" at the 1933 Century of Progress Exhibition in Chicago.[67]

Although significant technological challenges remained, a handful of steel framing and wall panel systems became commercially available by the late 1930s. Issues of condensation and corrosion continued to plague designers. While steel houses had significant thermal benefits, engineering the proper balance of thermal conductivity and thermal retention proved a difficult task. The higher cost of metal-fabricating equipment over wood also meant that steel-prefabricating companies would have higher up-front costs and larger investments in productive infrastructure. Recouping this investment required a reliable marketing and distribution system, which was further complicated by the uniqueness of steel housing. Firms often attempted to conform their designs to standard vernacular styles such as the ranch house or Cape Cod in an effort to increase market acceptance and deemphasize the "radical" nature of the house. Indeed, steel prefabricators trumpeted the "dual benefits" of attractive design and maintenance-free construction. The potential benefits of steel were alluring since it readily lent itself to large-scale fabrication. As a housing material, it provided product differentiation in terms of ease of maintenance, precision engineering, and permanence. Differentiation was a double-edged sword, however, since being different could be as much a problem as a benefit. Nevertheless, interest in steel prefabrication housing remained strong even though significant technological and marketing challenges remained.

Housing expert and industry observer Burnham Kelly noted that prefabrication attained the status of a "movement" during the 1930s. Indeed the general calls for improved technology and building processes of prior eras were given form and expression during the decade. Interest in government, industry, and the nonprofit sectors indicated that progress toward a more productive and rational system of housing was conceivably within reach. Yet the movement remained embryonic and uncoordinated. A wide range of interests recognized its potential significance but lacked the ability to integrate individual effort into a coherent whole. The enormity and complexity of America's housing problems made it difficult to assess the actual contribution that prefabrication could make. Additional time and experimentation would be needed, but the interest and energy of the movement demonstrated significant potential.

Ironically, the technological progress of the 1930s in some ways complicated the effort to define solutions. It created new possibilities but heightened the potential disruption of the status quo. Did the entrenched interests of the housing industry have nothing to fear but fear itself? Or would the prospect of new technologies replace the old power structure? Resting on decades of economic, political, and social tradition, the inherent irrationality and particularistic character of the housing industry could not be changed overnight. It is not surprising, then, that those who championed the transformational potential of prefabricated housing, whether in the name of profit, industrial recovery, or social benefit, encountered reluctance and resistance. The movement had achieved a certain momentum, yet its final destination remained unclear.

What was clear by the end of the 1930s, however, was the convergence of technologies toward new systems for low-cost residential construction. The development of new building components and materials fueled interest in developing overall systems that employed new products and processes. Working with a variety of materials, architects, engineers, builders, and manufacturers sought to understand how advancing housing technologies could be systematized to serve the mass market. Both government and industry had been active participants in the development of new technologies and processes, and in the coming years another type of convergence would emerge in the alignment of interests between industry and the federal government in the furtherance of housing technologies to serve the goal of providing affordable single-family housing in suburban settings to the middle-class market.

Two productive strategies emerged, each with important market and policy implications. One was the process of site prefabrication, a system that used an "outdoor assembly line" to construct single-family suburban subdivisions. The strict organization of labor and materials on a large scale made possible timely and cost-efficient construction. Specialized crews would rotate from site to site in a preset pattern. Once the foundation crew was finished, the framing crew took over, and the foundation crew moved on to the next site. Houses rose in a carefully orchestrated concert of production on street after street until the entire subdivision was finished. Site prefabrication balanced the requirements of traditional construction methods while imposing aspects of rationalized production. It was a flexible system that held the promise of providing average American families with decent suburban housing at an affordable cost. The power of uniting community planning with production efficiency came as close as anything had in the past to putting affordable single-family suburban housing within reach of the mass market.

The other approach was factory prefabrication, where an entire house was manufactured in a single factory for delivery to an assembly site. A far more challenging and potentially disruptive strategy, it held the promise of providing the full benefits of industrialized construction. Factory-made housing offered the greatest amount of control over the finished product since its sole focus was on the house. The abandonment of traditional construction processes meant that all aspects of a house could be subject to rigorous engineering on a systematic basis. Analysis, regimentation, and standardization were applied equally to every component of the house from floor to ceiling. The opportunity to control all aspects of the manufacturing process promised at last to bring the production and pricing achievements of the automotive industry to the housing market.

Operated on a large scale, factory-made housing offered a number of economic attractions. Unaffected by climatic or seasonal conditions, cost efficiencies inherent in continuous production could be realized. Further efficiency gained by pricing leverage on suppliers and the ability to influence supply chain quality control potentially gave factory-made housing competitive advantages in the marketplace. Indeed, at its most flexible, factory production could respond quickly to market conditions by introducing new features or new models based on market demand or preference. As in the automotive industry, the segmentation of product line to serve multiple price points meant that a firm could provide an entry-level house and retain loyalty as customers became more affluent and desired larger houses with more amenities.

Yet the apparent advantages and potential of factory-made housing came with enormous challenges. Capitalization costs would be vast. Attracting investors to an unproven and embryonic industry even in the best of economic times would require skillful planning and influential financial connections. Firms needed to "start big" to achieve production economies, and although there were plenty of examples of this, there were no models for the marketing and distribution of houses on a nationwide basis. Marketing and distribution structures had to support large-scale production as the franchised-dealer structure did in the automotive industry. Yet there was no evidence that the franchised-dealer structure—or any other distribution model—would work in the housing industry.

The key dilemma was that the separation of the house from the lot enabled the refinement of productive expertise but at the same time threatened to derail its benefits. The residential construction industry was based on a balance of production and realty interests spread across a variety of commercial and governmental entities. The completed house on a lot represented an accumulation of stakeholders, each accounting for only a small fraction of the total value of the property. Each stakeholder had an interest in increasing the economic value of its particular contribution with little regard to the whole. Prefabrication of any type, but especially factory-made housing, accumulated responsibility and influence over a greater proportion of the total value of the housing product. That any one entity could control a greater proportion of value opened up a world of possibilities. Unlike small builders or local developers, factory-made housing firms could justify expenses for research and development in the fabrication and assembly of the house itself and in the creation of nationwide marketing and distribution systems. As Burnham Kelly observed, "there has been no element in the house building industry with sufficient motivation and with sufficient power and means of control to initiate fundamental changes." Large-scale factory-made housing might be able to capture a sufficient portion of the total value equation to implement systemic changes in the housing market.

The prospect of such a disruption, however, raised numerous issues. The private market remained wary of industrialized housing, largely because they were uncertain about the ultimate impact on their industry. Prefabricators thus found it difficult to predict where opposition would arise since so many different entrenched constituencies were affected by new processes. A combination of active and passive resistance emerged throughout the industry. Passive resistance was far more dangerous to the development of industrialized housing since the prefabricated housing industry had to change the system to become truly effective, while entrenched interests had only to maintain the status quo. The entire American system of housing was designed to protect these interests. Financing practices, building codes, labor systems, land tenure, and real estate sales practices presented numerous opportunities to impede the development of industrialized housing. By the end of the 1930s, it was clear that the private housing market would not and could not provide the energy and resources to drive the development of industrialized housing. A different type of convergence would be necessary.

The unprecedented involvement of the federal government in the housing market during the depression era, and particularly its interest in housing technologies, fueled a movement toward an alignment of interests between the federal government and the prefabricated housing industry in the furtherance of public policy goals. Faced with the continued intransigence of the private housing market, the federal government was increasingly willing to support prefabricated housing to achieve societal goals—first for defense housing during World War II, then as a potential solution to providing single-family suburban-based housing for the postwar middle-class market. The embrace of productive technology in the housing industry was a logical extension of government support for productive technologies in defense-related sectors and drew heavily on the paradigm of government support to industry that transformed the American economy from depression era malaise to the world's premier economic power in a few short years.

With war clouds on the horizon, supporters of industrialized housing saw an opportunity to advance the industry as a potentially powerful weapon in the cause of housing reform. After years of speculation and promise, the technology and the will to invest in it appeared likely to converge during the war years. There were many positive signs that prefabrication would change the housing system for the better. The interest of big business was significant in itself, for it demonstrated the viability of the industrializing paradigm. Even though initial efforts had been unremarkable and driven by the desperation of the depression, technical advancements were made and a greater understanding of systems engineering had been achieved. Greater appreciation of the entire range of implications for prefabrication was also evident, especially the importance of marketing, promotion, distribution, and setting. Using the factory system to produce housing for the American mass market seemed innately rational. How best to achieve it was not known, yet in the crucible of wartime crisis innovative and influential approaches would emerge as the government, the housing industry, and various reform constituencies sought to further opportunities for better housing.

CHAPTER TWO

Crisis Management and the Rise of the Lustron Corporation

War is the health of the machine.
—Lewis Mumford, *Technics and Civilization*

As the United States prepared for war, the prefabricated housing industry faced an exceptional opportunity to align its development with the numerous federal housing programs associated with defense mobilization. The industry had taken important technological and organizational steps during the 1930s yet remained a peripheral development with uncertain commercial promise. It was far more influential in theory than in practice, although conventional builders and housing-component manufacturers had to some degree embraced the movement's emphasis on standardization. Yet the pressing need for housing to serve the massive internal migration accompanying military mobilization promised to accelerate the development of integrated building technologies and assist innovative approaches to the design and construction of residential housing.

Like entrepreneurs in other industries affected by the mobilization effort, leaders of the long-depressed housing industry welcomed the economic opportunities presented by the defense program. Prefabricated housing firms especially sought to leverage their involvement to gain market advantage and political influence at the expense of conventional builders. As in the early years of the depression, the social and economic disruption of wartime presented opportunities for experimentation and weakened the ability of entrenched interests to oppose new ideas. The scale, scope, and persistence of the wartime housing crisis gave rise to partnerships between prefabricators and government agencies, which threatened the particularistic conservatism of the private housing market. How long this window of opportunity would remain open was unclear, but it was evident that the defense housing program's support for aggressive, innovative thinking about housing design and production raised suspicions that longstanding economic and political relationships within the industry could be altered by renewed federal activism. The prospect of better living through better housing after the war also became an important influence on the advancement of housing technologies during the war. "After total war can come total living," one wartime advertisement predicted, and as the war progressed expectations for new lifestyles based on home ownership increasingly seemed a logical and just recompense for service to God and country.

Production for Use

With war clouds gathering and the industrial mobilization program gaining momentum, the Roosevelt administration soon faced the thorny issue of housing and its connection to worker productivity. As in the First World War, defense workers needed adequate housing, and they needed it quickly. The administration's strategic objective of dispersing American war production across the nation and on the periphery of established cities heightened the crisis. New defense plants on the suburban fringe suited production efficiency but left workers far from established neighborhoods. The reconfiguration of the suburban periphery to house defense workers marked the beginning of a powerful transformation. Housing interests caught on quickly, anxious to share in the creation of a new setting where production and planning intersected with patriotism and profit.[1]

American families drawn to the economic opportunity of the new defense centers found no suburban utopia in the early stages of the mobilization effort. Their elation at the prospect of steady work and upward mobility after many years of depression was soon mitigated by the realization that they had nowhere to live. Some turned around and went home, others lived in their cars or sought shelter in abandoned trailers, grain bins, or railroad cars. In Bremerton, Washington, workers employed at a local naval yard inquired about the possibility of lodging at the city's jail. The lack of decent shelter and basic sanitary services threatened the viability of the defense program and quickly captured the administration's attention. Not since the early days of the New Deal had the federal government focused so keenly on housing, and its response reflected earlier tactics. It first sought to stabilize the private housing market and spur construction by providing subsidies for the construction of defense housing and increasing opportunities for defense workers to purchase homes. Working with the Defense Homes Corporation (DHC), an agency created to underwrite construction loans to developers, and the Federal Housing Administration (FHA), the administration encouraged defense workers to abandon renting and purchase new homes. In 1940 the FHA initiated a program to promote home ownership among families with at least $2,500 annual income by guaranteeing a greater percentage of mortgage loans. With as little as $25 per month, a family could purchase "a modern home—comfortable to live in, attractive to look at, convenient to pay for."[2]

Indeed, these programs increased home ownership, and for families who could afford to purchase a home they were a godsend. However, as the dimensions of the mobilization effort became clearer, the administration recognized that a far more comprehensive program would be necessary. Even with generous subsidies, the private housing market could not produce defense housing quickly enough and in sufficient quantities to sate the demand of the nation's ever expanding war production machine. In September 1940 Representative Fritz Lanham introduced legislation providing for federal construction of defense housing. Supported by a broad coalition of labor and social welfare groups, construction and real estate interests, and the home finance industry, President Roosevelt signed the National Defense Housing Act, popularly known as the Lanham Act, in October 1940. An opponent of low-income public housing, Lanham protected the private housing market by stipulating that federally assisted defense housing would only be built in areas ill served by private interests. He later secured an amendment that prohibited the conversion of defense housing to low-income public housing without congressional approval.[3]

The Lanham Act authorized the Federal Works Agency (FWA) to construct housing for "persons engaged in national defense activities" and provided ample budgetary flexibility to experiment with a broad range of construction techniques and housing technologies. FWA director John Carmody, a veteran New Dealer and an architect of rural electrification, saw an opportunity to create model communities that would influence the course of residential construction and planning for decades to come. Carmody used the FWA's constituent agencies, the U.S. Housing Authority (USHA), the Defense Housing Division (DHD), and the Mutual Ownership Defense Housing Division (MODHD), to advance housing through research and development in building materials and systems, architectural innovation, and financial incentives. Another federal agency active in defense housing during the early stages of mobilization was the Farm Security Administration (FSA), which had a strong record of constructing emergency housing and working with innovative construction systems, including prefabrication.[4]

From the signing of the Lanham Act in October 1940 through early 1942, the federal government built 207 defense housing projects and had 200 more in various stages of planning and construction. A review of several prominent examples reveals the government's consistent interest in innovative architecture and improved building processes. The USHA, DHD, and FSA recruited noted architects and supported research and experimentation with new building technologies in the field and in the laboratory. Enlisting a modernist approach to architecture and design, the federal programs solidified the connection between emerging building technologies and the desire for new residential forms. Open floor plans, efficient use of space, and greater integration of the outdoors through the use of larger windows and outdoor living spaces reflected a union of technology and design previously unknown on such a large scale.[5]

The defense housing programs influenced postwar residential design in profound ways, both in terms of architectural form and production processes. As architectural critic Peter S. Reed has observed, "the architectural triumphs and design innovations surrounding World War II . . . marked a turning point—thereafter modern architecture had come of age, emerging after the war as a hallmark of democracy, capitalism, and the Establishment."[6] The integrative approach that had long been sought by various constituents of the American housing system seemed at hand. Accelerated and concentrated by the defense housing crisis, the financial and organizational resources of an activist government advanced the cause of industrialized housing more in two years than in the previous four decades. The architectural profession was a key beneficiary and influential force within the defense housing program. The guiding hand of the architect played a role in fostering a productive revolution in housing while creating aesthetically attractive and socially rich communities. Apparently, systematization, standardization, and speed did not have to result in "dreary" housing but could be imaginatively employed to produce a new vision for American residential life.

The DHD under the leadership of Clark Foreman played a key role by enlisting a number of well-known architects. Foreman was sensitive to the mediating influence that architects exerted between production technologies and social and aesthetic considerations. He also hoped that associating the DHD with noted architects would mitigate any stigma associated with "government housing" and increase public acceptance of his program. Widely acclaimed in the press, the program attracted architects William Wurster, Walter Gropius, Eero Saarinen, Marcel Breuer, Louis Kahn, Alfred Kastner, Richard Neutra, and Frank Lloyd Wright. The program issued strict guidelines in accordance with the Lanham Act, which in 1940 allocated $150 million for defense housing and limited construction costs to $3,500 per dwelling unit. Despite these financial limitations, DHD architects produced a number of innovative designs.[7]

A notable example came from architects Walter Gropius and Marcel Breuer, well-known modernists and members of the German Bauhaus, one of the most influential architecture and design schools of the twentieth century. Gropius had gained considerable experience building low-cost housing in Germany before immigrating to the United States in the 1930s. The DHD commissioned Gropius and Breuer to design Aluminum City Terrace in New Kensington, Pennsylvania, near Pittsburgh, for workers at an ALCOA plant. The project's design featured many aspects of modernism such as ribbon windows, flat roofs, and open floor plans, even to the point of excluding walls between the bedrooms. Ironically, many of these features were typical in the very expensive homes designed by Gropius in the late 1930s. While some tenants did find the exteriors and interiors severe, most were happy with the large windows and open interior plans.[8]

Frank Lloyd Wright received his first government commission from the DHD to design a 100-unit housing project in Pittsfield, Massachusetts. With his usual flair, Wright proposed a quadruple, or "cloverleaf," house design consisting of four units that shared central walls. Ever modest, Wright described the design as "standardization complete without stultification or loss of individual freedom." The innovative and attractive cloverleaf design met the program's $3,500 unit price ceiling, but just as Wright completed the construction documents for the project, local political turmoil intervened and the project never left the drawing board.[9]

In late 1941 the DHD hired Richard Neutra to design a defense housing community in San Pedro, California, near several large shipyards. Known as Channel Heights, the 165-acre site was dramatic, including irregular topography and stunning views of the Pacific Ocean. Neutra staggered dwellings for 600 families throughout the site with a geographic harmony that afforded most residents an ocean view without sacrificing privacy. Horizontally sliding windows and a combination living/dining area contributed to a feeling of interior spaciousness. The project also featured a meandering main road and a park with pedestrian paths that tunneled under the road at various points and extended through the entire development.[10]

Led by innovative architects, the defense housing program produced a number of excellent housing developments that served as models for the integration of technology and community. However, notwithstanding these efforts, by the end of 1941 the housing program was the furthest behind schedule of all defense building efforts. A change in policy was in the wind, and the administration began to consider the efficacy of the defense housing program's concern for community building. Particularly influential was the work of architects William Wurster and Vernon Demars, who had separately designed dormitories, cabins, community buildings, and single-family houses for several thousand workers and their families in Vallejo, California, near the San Francisco shipyards. Half of Wurster's one-story houses were constructed of stressed-skin, glued plywood panels and the remaining half with Homosote, a new wallboard made of wood pulp and ground newspaper bound with resin. The houses used framing systems that eliminated the need for load-bearing interior partitions, producing open interiors with large expanses of glass. Panels were trucked to the site, and walls were preassembled on the concrete slab floor and raised into place. Significantly, the structures at Vallejo were "demountable," meaning that they could be easily torn down or relocated.[11]

After Pearl Harbor and the declarations of war, the administration was anxious to remove all delays and bottlenecks in what was now the war housing program. In February 1942 Roosevelt signed Executive Order 9070, establishing the National Housing Agency (NHA) and transferring to it responsibility for sixteen constituent agencies of the federal housing enterprise. Headed by John Blandford, the NHA coordinated all aspects of housing production including building materials, labor, construction methods, land acquisition, and financing. Blandford adopted a more rigorous commercial approach to the defense housing crisis, eschewing any aspirations for social reform as evidenced in previous programs under Carmody and his associates. Carmody had resigned the previous October for health reasons, and Blandford soon replaced Carmody's senior staff with his own pragmatic, production-oriented team.[12]

Blandford's appointment signaled a new era in the war housing saga, one in which the prefabricated housing industry would play a greater role. The focus on production signaled a departure from the construction of permanent communities that could serve as blueprints for postwar residential housing developments. Instead, the NHA sought to provide adequate but minimal shelter for war workers that could be "demounted" after the war. Supported by real estate interests and conventional homebuilders, the plan addressed fears that permanent government housing would compete against their interests and drive down property values in areas attractive for future development. For the prefabricated housing industry, the change in policy presented a tremendous opportunity. It was confident that it could deliver the type of housing that the government needed while at the same time remain shielded from competition and criticism from conventional builders who were willing to cede this part of the housing market for the duration of the war. Under cover of federal largess, prefabricators hoped to exploit their position against conventional builders by using the opportunity to refine productive technologies and use profits to fund research into marketplace barriers.

The reorientation of federal policy from the "utopian schemes" of the initial phase of the defense housing program to an ardent focus on production was more than a fortuitous development for the prefabricated housing industry. It meant that the government's interest in housing technologies was no longer associated with the broad range of housing reform and social enterprise that had influenced earlier efforts, including many of the notable New Deal housing achievements. Technological innovation would play an important role, but not in the sense that the government would compel the private market to alter its modus operandi. Government support for prefabrication and other housing technologies would be limited to crisis conditions and would not produce permanent housing that competed with the private market or devalued areas for future residential development. The administration had made its peace with the private housing market and reinforced the idea that the government would continue to provide financial and other infrastructure incentives to foster the expansion of home ownership but would not interfere in the future settlement of suburbia initiated by the mobilization program.

Even so, the private housing industry had to contend with unprecedented governmental involvement in the market, a fact that irritated it philosophically and politically far beyond the actual economic consequences for potential lost revenue due to government competition. The industry wondered how to convince the American public that the federal government should retreat from its activist stance once the war was over—without, of course, retreating from its indirect subsidies and market stabilization efforts that supported private development. One strategy was to reinforce in the public mind a differentiation of quality between government-built and privately built defense housing. This was largely a smoke screen since all defense housing was a semipublic enterprise in one way or another. Yet they claimed that government-sponsored housing was flimsy and unlikely to last the duration of the war. Indeed, not all defense housing projects were desirable or even livable by today's standards, but this stance ignored many innovative and attractive community projects like Channel Heights. To promote their views, industry leaders established the National Association of Home Builders (NAHB). NAHB quickly became an extremely powerful organization and reflected a broad industry perspective. It frequently and vociferously opposed all government-sponsored housing initiatives that it deemed detrimental to its interests. Important constituent groups of the NAHB included the U.S. Savings and Loan League, the National Association of Retail Lumber Dealers, and the Mortgage Bankers Association.[13]

The prefabricated housing industry stood in the midst of these developments, anxious to capitalize on the reorientation of the war housing program but uneasy in its relationship with conventional interests. Leveraging government opportunity into a stronger competitive position in the private market was a path fraught with peril. Moving beyond the "temporary" market into direct competition in the traditional marketplace, where numerous barriers existed in the form of building codes and localized financial and real estate practices, meant competing on uneven and politically hazardous ground. Yet the prospect of spearheading a mass-market production paradigm presented an opportunity to accelerate development under governmental aegis. With the federal government as a "sponsor," the industry felt sure that it could stake a claim to a profitable share of the residential construction market. In the meantime, prefabricated housing firms embraced the new direction of federal war housing policy with great energy and imagination.

Historically strong on production technology but lacking marketing savvy, prefabricated housing firms looked to the government to guarantee a profitable market and thus remove one of the most vexing problems inhibiting commercial development. In 1940 there were only thirty prefabricated housing firms producing on a steady basis. The output of the industry was miniscule, accounting for less than 1 percent of all the single-family houses built in the United States during that year.[14] The industry was clearly on trial during the early stages of war mobilization. Many in government remained skeptical of its ability to produce quality products in a timely manner on a large scale. The demonstration period of the 1930s was over, and the industry now had to step to the fore with attractive products to sustain government investment. However, prefabricators anticipated garnering an increasing share of the defense housing market by increasing the speed of production and emphasizing cost savings.

As the American war production machine expanded and the need for worker housing increased, Blandford and the NHA increased pressure on the industry to provide more houses in less time. The industry responded quickly, growing to over one hundred firms by early 1943. Entrepreneurs like Foster Gunnison of New Albany, Indiana, whose firm pioneered new assembly line production techniques, advanced the slogan "Press a Button, Get a House," which reinforced the idea that the industry was ready to deliver on the promise of mass-produced housing. The "production focus" clearly indicated a lessening of concern for integrating prefabrication with the community development approach championed by Carmody, which employed prefabrication but focused attention on environmental and communal factors.

Temporarily freed from concerns over the "proper" placement of its houses in neighborhood settings, prefabricated housing firms focused their attention on refining production technologies and engineering expertise. Companies working with wood systems made incredible progress in a short period of time, advancing the quality of wood-based prefabricated housing and exerting considerable influence over conventional building practices as well. Stressed-skin plywood panels and wood sandwich wall systems using various combinations of products came of age during the war. Modular design became much more prevalent as engineers focused on production flexibility as well as high-volume production. The modular concept employed in a variety of commercial applications such as war plants helped prefabricators and architects conceptualize the residential construction process in a more efficient and holistic manner. The freedom to design "houses" rather than "communities," coupled with the industrial focus of the war housing program, unleashed creative energies that had far reaching implications. A key development was the use of modular components, such as bathroom or mechanical systems, which advanced standardization across the entire construction industry and helped all types of residential builders cut costs and improve production.[15]

The prefabricated housing industry also gained an increasingly powerful sense of self-awareness during the war. Established in 1942, the Prefabricated Home

Manufacturers Institute (PHMI) became the primary trade association for the industry. The Institute established quality standards, assisted startup firms with cost and accounting studies, and encouraged members to identify and address nagging problems such as marketing and distribution. PHMI functioned as a clearing house for information and fostered discussion of internal and external issues affecting the industry. It focused a great deal of attention on public relations, especially concerning issues of durability and design that presented important barriers to public acceptance. In 1943 the Institute began publishing *Prefabricated Homes,* a monthly magazine chronicling all facets of the industry. Aware of the industry's intimate connection to the government, PHMI also initiated a lobbying campaign to influence national and local governmental policy. From encouraging additional federal support to assisting the revision and adaptation of local building codes, PHMI became an active and influential voice for the industry.[16]

The war years accelerated the maturation of prefabrication in a number of important ways. The productive orientation of the war housing program forced the industry to bring its product to the market in a timely manner. Indeed, the industry accounted for 200,000 of a total 1.6 million houses built under the various programs. The "houses of the future" orientation of the industry was replaced with a more pragmatic and commercially savvy approach. The war brought a needed dose of fiduciary reality and forced companies to face the range of thorny issues surrounding the prefabrication process. Even though the product that emerged was anything but fully realized, it was a start. The "dreamers to doers" pressure of the war also produced a sense of optimism that prefabrication's golden age lay just around the corner in the postwar era—a time where process, product, and setting would be fully industrialized and integrated.[17]

The war housing program gave the industry an opportunity to "practice" developing simple houses in large volumes. Lessons learned in improving engineering and manufacturing methods, as well as site erection efficiency, provided keys to serving the mass market. A crucial lesson in scale was also learned at a minimal cost to the industry. Production volume matters in industrialized building, and its relationship to unit cost must be intricately understood and accounted for before any plans to serve the mass market with affordable single-family homes can be implemented. The positive influence of wartime building on the overall financial health of the industry and its ability to attract investment, obtain credit, initiate research and development, and expand operations proved invaluable as well. Prefabrication's influence on building processes in general was also accelerated by the war. Large conventionally built war housing projects employed more effective techniques of prefabrication, such as materials standardization, specialization of labor, and rational planning of material flow. As much as possible, traditional builders brought factory methods to the site in the form of power tools, jigs, cranes, and other material handling equipment. Site prefabrication was indeed a major beneficiary of the defense housing program's focus on production efficiency. The government thus played an important educational role for the merchant builders who would construct most of the postwar suburban housing in the United States

Yet the industry was hardly home free. Numerous challenges and uncertainties remained. The focus on production in the service of the war housing crisis mitigated interest in marketing and distribution. Serving one customer rather than the open market fostered growth but impeded consideration of how the industry would serve the particularistic and ever changing preferences of the home-buying public. The perception of quality and permanence was crucial in attracting interest among postwar consumers, and the industry had not fully processed the marketing implications of producing its demountable, "bare-bones" models during the war. PHMI and other industry groups began to research and analyze public perceptions of prefabricated housing and devise methods for addressing the inherent tensions between standardization and design flexibility.[18]

The public had strong personal preferences about housing forms nurtured by decades of tradition. Could prefabrication evolve to serve their wide ranging and often fickle tastes? The issue touched all phases of the housing equation. Americans were ready for the "house of tomorrow" today, but would they recognize it if they saw it? Conventional builders played on the association between prefabrication and impermanence, and organizations such as the PHMI worked diligently to change public opinion. It would prove a difficult task. In late 1944 the Curtis Publishing Company conducted a poll revealing that while 74 percent of respondents had heard of prefabricated housing, only 17 percent would consider living in one. A subsequent *Fortune* poll reported that a majority of respondents believed that prefabricated houses were not soundly constructed, lacked individuality, and would not last more than a few years.[19] Although the result of the defense housing program's focus on quick, simple, and demountable housing, the polling was nevertheless disconcerting. What fed the industry's growth also cost it in terms of future viability outside of crisis conditions. These concerns notwithstanding, networks of opportunity presented to the prefabricated housing industry during the war had positioned it for further growth.

The Veterans' Housing Opportunity

As the American war machine marched ever closer to victory, concerns over the transition to a peacetime economy gained widespread attention. Would the depression return? Should wartime controls be extended or curtailed? Would people return to cities when the suburban defense plants scaled back? Nobody knew for certain. The promise of postwar prosperity and heightened desires for the good life pervaded the popular media, but significant numbers of articles also warned of the economic uncertainties inherent in the reconversion process. The housing industry, as always, was a key focus of concern for a wide range of political, economic, and social observers and actors. The unique blend of commerce and community inherent in the American housing system again took center stage in the national arena.

As a result of its role in the war housing program, prefabrication became a more visible and viable element in the planning and implementation of postwar housing strategies. The industry's consistent presence in federal policy discourses and actions as well as the continuing specter of its disruptive potential in the private market demonstrated it newfound gravity. Social, economic, and political forces would impact the industry in the immediate postwar era in profound ways, shaping its future development as they had in the past. However, unlike the 1930s or the war years, one company—the Lustron Corporation—would come to represent the fortunes of the entire prefabricated industry.

The dream of home ownership in a suburban setting, which had been cultivated by the press and by the government since the 1920s, had taken root. Indeed it was no longer a dream, but an expectation, and returning veterans expected and demanded that the government they had served so well do as much as possible to make good on its promises. Interestingly, according to a poll in the *Saturday Evening Post,* 86 percent of Americans wanted new houses, not apartments or "used" houses.[20] Reflecting years of pent-up demand and memories of substandard housing during depression and war, returning veterans and their families wanted new homes with all the conveniences that modern industry could provide. The passage in 1944 of the Serviceman's Readjustment Act was an important first step. The GI Bill of Rights included two powerful and transformational elements that would define middle-class life in the postwar era—free college tuition and interest-free mortgages that required little or no money down at the time of purchase. Government underwriting of veterans' mortgages provided a financial foundation for the expansion of home ownership. All veterans needed now were the houses.

With the right to decent housing affirmed by federal actions during the depression and war, most Americans looked to the federal government to extend that commitment in the postwar era. The GI Bill was an excellent start, but the government also needed to facilitate housing production to satisfy a potentially enormous demand. Few in government had a clear picture of the dimensions of the problem, and many thoughtful Americans believed that a return to depression was almost certain. In general, the Truman administration initially underestimated the impact of what would come to be known as the veterans' housing crisis. The war ended suddenly and caught the administration without a coherent strategy and with divergent views in its own ranks. Lacking clear definition, Truman's housing policies produced bickering, posturing, and intimidation within the administration and muddied relations with Congress on a range of reconversion issues. How best to stimulate the production of housing—not just temporary housing, but the kind of single-family suburban-based "new" housing that most American desired—was a complex and nuanced problem that called upon numerous constituencies to craft commercial and communal strategies to meet the crisis.[21]

Approximately ten million veterans were discharged from the U.S. armed forces during 1945 and 1946. Throughout the nation millions of reunited families and newly married couples sought shelter with relatives or occupied a variety of other "residences," including garages, attics, boxcars, and even chicken coops.[22] Veterans' groups led the nation in a powerful and emotionally charged call for governmental action. The challenge to the prefabricated housing industry, and also an important source of its appeal to government policy makers, was the potential ability to unite production with form and setting. This was the brass ring of prefabrication—the industrialization of residential construction accomplished within established norms of community development. A technological and production focus was a key result of the second phase of the war housing program, and it had fostered creative thinking across the industry as entrepreneurs calculated how best to capitalize on new methods and materials once the war was over. Support for prefabrication in the postwar era was thus in keeping with governmental policy. The question remained, however, if prefabrication held the promise of a better postwar America by providing affordable housing for the wage earner market segment that had been long neglected by conventional builders.

As the flow of discharged veterans increased, pressure mounted quickly on the administration to address the production issue. Fears of returning depression clearly heightened interest in industrialized housing as a pathway to sustained prosperity and ongoing social investment. Direct and indirect investment through the use of government agencies such as the NHA and the VA and government corporations such as the RFC might provide needed capital and credit to foster integrated production as well as allocate needed materials to launch an industrial revolution in the housing industry. However, complicating the response of the Truman administration to the veterans' housing crisis was a split between two factions. Truman himself was ideologically inclined to end price controls on housing and production restrictions on building materials, but others believed that the continuation of controls could soften the impact of any potential economic decline and grant the administration more control over the reconversion process.[23]

A leading advocate for ending controls was Julius Krug, head of the War

Production Board (WPB). John Snyder of the Office of War Mobilization and Reconversion (OWMR) supported Krug and agreed with the NAHB and other representatives of the private housing industry that further controls would hamper reconversion. Chester Bowles, head of the Office of Price Administration (OPA), strongly opposed that viewpoint. Bowles feared disordered production, price gouging, and widespread shortages of affordably priced housing. Although the NHA and organized labor sided with Bowles, Truman ended controls and restrictions on 15 October 1945 in the hopes that renewed construction activity would spur the economy.[24]

A building boom indeed resulted. However, it did not produce affordable housing for veterans but a "rash of race tracks, mansions, summer resorts, bowling alleys, stores, and cocktail bars." The administration quickly recognized its mistake. Bowles had continued to lobby for renewed controls, using an appeal to veterans' rights as his main argument and pointing out that promoting high-market building was inconsistent with the goals of the GI Bill. In November, Bowles turned to his friend Congressman Wright Patman, the liberal Democratic chairman of the Banking and Currency Committee, who introduced a housing bill that gave the executive broad powers to reinstate controls, stimulate production, and create new agencies to further the cause of affordable housing.[25]

Truman used the bill to refocus the administration's energies on an emergency housing production program. In December he announced support for production priorities and housing price controls and began formulating a comprehensive housing program. In January 1946 the president announced the creation of the Office of the Housing Expediter to lead, coordinate, and facilitate the government's response to the crisis. He appointed Wilson Wyatt, a liberal Democrat, attorney, and former mayor of Louisville, Kentucky, to head the office. Truman also named Wyatt administrator of the NHA. As mayor of Louisville, Wyatt had attracted national attention as a successful reformer and consensus builder. Although he lacked direct experience or expertise in housing, the president believed that his administrative abilities and forceful personality well suited the demands of the position.[26]

Truman instructed the young and energetic administrator to "make no little plans" and conferred sweeping powers to "search out bottlenecks at whatever level of industry or of Government . . . and make the machinery of housing production run as smoothly as possible."[27] Wyatt faced a double-edged sword. If he moved too boldly, he risked engendering bureaucratic intransigence and partisan maneuvering, which heightened the possibility of a grand and very public failure. Choosing a moderate path also held dangers, for the administration did not wish to incur the wrath of powerful veterans' groups who demanded an immediate solution to the crisis. Though cautious in his first public statements, Wyatt nevertheless promised to "lick" the housing crisis "in the not too distant future."[28]

Defining goals and developing strategies to meet the crisis occupied Wyatt's first weeks in office. He immersed himself in housing literature and held a series of conferences with government leaders, veterans' groups, labor unions, real estate associations, and building materials producers in an effort to gain a better understanding of their perspectives on the housing crisis. This "cram course," as he described it, was necessary to "learn where we were, where we wanted to go, and how we might get there." For Wyatt this was a crusade—a continuation of wartime efforts and an extension of the same spirit of mutual interest and cooperation between industry and government that had won the war. Comparing housing production for returning soldiers to wartime production for soldiers on the front lines, Wyatt tapped into powerful feelings for America's returning heroes and aspirations for their domestic welfare in the postwar era.[29]

After five weeks of study, Wyatt presented his Veterans' Emergency Housing Program (VEHP) to the American public on 7 February 1946. Truman issued a statement accompanying the report that praised the program as "bold, vigorous, and eminently practical" and guaranteed "the complete and unqualified support of the Administration."[30] Wyatt's program reflected his belief that the housing shortage could only be relieved by the continuation of strict governmental economic controls. Without controls, he argued, builders would produce more profitable higher-priced homes that few veterans could afford. Wyatt defined affordable housing as new houses costing $6,000 or less or renting for under $50 per month. Mindful of the postwar desire to end wartime restrictions, Wyatt nevertheless called for sacrifice from all sectors of the economy: "Neither business-as-usual, labor-as-usual, building-as-usual, nor Government-as-usual will suffice."[31]

Wyatt established the lofty goal of constructing 2.7 million low- and moderate-cost houses by the end of 1947 and proposed a number of emergency measures to achieve that goal. These included giving preferences for veterans and their families in the rental or purchase of homes, tripling the present labor force engaged in residential construction, postponing all nonessential construction for the remainder of the year, and providing federal cooperation and assistance in the development of home sites. Additionally, the formation of "emergency housing committees" in cities throughout the nation would serve as a link with the federal government to assess local needs and monitor compliance with the program.[32]

To encourage the production of innovative building materials, Wyatt advocated the use of market guarantees, rapid tax amortization for new or converted factories producing essential building materials, the allocation of surplus war plants to increase production capacity, and federal assumption of "undue economic risk" for research and development into new approaches to residential construction. Support for all types of prefabricated housing provided a key component in the VEHP.

Using federal assistance in the form of loans from the RFC and market guarantees from the NHA, Wyatt believed that prefabrication could account for as many as 600,000 of the projected 2.7 million units.[33]

The house-starved American public overwhelmingly supported the VEHP. According to one of Wyatt's associates, in the weeks after its announcement a wide range of interests "seemed to vie with one another in praising it."[34] Never had a housing program attracted so much favorable public attention and comment, a development that brought even more pressure to bear on the VEHP. Wyatt recognized a good thing when he saw it and leveraged the enthusiasm to further his program. However, because the VEHP derived its power from the Second War Powers Act, due to expire in June 1946, the administration needed congressional approval to implement the full range of Wyatt's plans. Some conservative senators and representatives expressed concern over what they called "the socialist provisions" of the program, such as continued controls on material allocation and financial support for the prefabricated housing industry. Arguing that an unrestricted economy would produce housing quicker and more efficiently than a new bureaucracy, conservatives charged that Wyatt possessed dictatorial powers to favor any manufacturer he wished.[35] Despite these reservations, the Veterans' Emergency Housing Act (VEHA) passed both houses of Congress and Truman signed it on 22 May 1946.[36]

The Act represented a triumph for the administration and an endorsement of Wilson Wyatt. Full of confidence, he announced that his program would now "go into high gear."[37] Following congressional approval, Wyatt moved quickly, using his powers to divert scarce construction materials into housing for veterans and curtailing all "nonessential" construction.[38] Although he believed that the supply of materials presented the greatest barrier to the success of his program, he failed to anticipate the unswerving resistance of the housing industry to the continuation of governmental controls, as well as the crippling effects of labor disputes in key industries that slowed the production and distribution of essential materials. Wyatt stood helpless as shortages of gypsum board, plywood, lath, bricks, pipe, and nails continued. Even with his broad powers, he could not create instant supplies or stop strikes against key producers.[39]

Despite these obstacles Wyatt remained optimistic and reported in June that the program was well on its way toward achieving the goal of 1.2 million housing "starts" during 1946.[40] However, by late summer the construction industry had started only 600,000 units and completed fewer than 190,000 permanent dwellings. Public opinion, led by veterans' groups, began to turn against the VEHP.[41] Facing extreme political pressure and realizing that conventional building would never meet his goals, Wyatt attempted to save the VEHP and his Washington career by reorienting the program toward the development of prefabricated housing. The enthusiasm and vigor with which Wyatt pursued production through prefabrication reflected a genuine belief in the promise of the industry as well as a desire to spur conventional builders to act with greater speed lest they be left behind. Wyatt's strategy was rationalized by his acceptance of the belief, best articulated by General Motors' Chairman Alfred P. Sloan, that prefabricated housing "could well be the greatest industrial opportunity since the development of the automobile."[42]

Enter Strandlund

There was another man in Washington who felt the same way—Carl G. Strandlund. Strandlund was vice president and general manager of the Chicago Vitreous Enamel Products Company, one of the leading firms in the American enameling industry. The company had dispatched him to Washington to seek an allocation of steel for the company's fledgling enameled-steel panel construction business. A seasoned mechanical engineer with an entrepreneurial zeal and zest for life that attracted both friends and attention in great measure, Strandlund was at the pinnacle of an impressive career. Portly and balding and in his mid-fifties, he possessed seemingly inexhaustible stores of energy and an impressive array of skills and abilities. Soon he would need all his talents, for his trip initiated a chain of events that united the technological, entrepreneurial, and political elements that gave rise to the Lustron Corporation.

Strandlund's easy demeanor and engaging, polished personality had facilitated his entrée into a wide array of social and commercial circles. He was active in industry associations and social clubs but above all loved sports, especially horse racing. The sporting life of the race track held a special appeal for Strandlund, for it fed his risk-taking nature and boisterous spirit. He owned and raced a series of thoroughbreds at Arlington, Hialeah, and Pimlico. Strandlund enjoyed wagering on the ponies, and even though he lost more than he won he shrugged off his losses and moved on to the next prospect. Money was a means to Strandlund, not an end. He firmly believed that he could always make more.[43]

A man of considerable style, Strandlund also possessed the practical, ordered mind of an engineer, keen commercial instincts, and a lust for innovation. It ran in his veins. Strandlund was born in Sweden in 1888, the grandson of one of that nation's most prominent engineers. When he was four, his family immigrated to the United States and eventually settled in Moline, Illinois. His father found work at John Deere and Company and eventually received over three hundred patents for improvements to farm machinery. Carl showed an early aptitude for "hands-on" engineering, spending countless hours beside his father in the family's workshop. He understood the most mundane technical details and could explain them to virtually any audience.

Carl Strandlund, Lustron's founder and president. Photo Credit: KDN Videoworks

He was also adept at motivating people—a leader who could inspire employees to achieve beyond their expectations and a visionary who could sway customers and influence investors to embrace his ideas. People seemed to accept him at a fundamental level, and he reciprocated by infusing enthusiasm and excitement into every endeavor.[44]

After graduating from Moline High School, Strandlund studied engineering through correspondence courses. His first job was with the Minneapolis Moline Power Implement Company, where he worked from 1923 to 1935 as a production engineer. Here Strandlund's reputation as a "guy who could design for production" was born. During his tenure at Minneapolis Moline, he amassed over 150 patents, ranging from improving tractor seat design, significantly increasing weight to horsepower ratios in tractors and combines, to using rubber tires instead of metal wheels on tractors—all of which increased the market potential and consumer acceptance of the company's products.[45]

From Minneapolis Moline, Strandlund joined the Oliver Farm Equipment Company as director of engineering and development. He headed manufacturing operations at Oliver's four plants in Iowa, Indiana, Michigan, and Ohio. His focus on production efficiency and supplier management transformed operations and in 1937 he became president of the firm. As president Strandlund increased revenues from $20 to $120 million. Before joining Chicago Vitreous, Strandlund worked as a consulting engineer for the Chicago Steel Foundry Company and for B. F. Goodrich in Akron, Ohio. His work at Goodrich on improved designs for tank treads drew the attention of the War Department and gave him his first exposure to the intricacies of the defense contracting business.[46]

In April 1942 Strandlund joined Chicago Vitreous with specific duties to oversee its conversion to war production. The company had secured contracts from the ordnance department of the Army for the production of gun turrets, armor plating for tanks, and, intriguingly, powdered eggs for ration packets. Chicago Vitreous engineers quickly mastered powdered egg production by converting machinery formerly used to grind enameling clays.[47] However, the production of gun turrets and armor plating required expertise beyond the company's capabilities. Strandlund's outgoing personality, quick wit, and production expertise immediately impressed the owners of Chicago Vitreous, brothers William and Emanuel Hogenson. The enthusiastic sportsman with an air of easy opulence fit well into the corporate culture, and his hand-tailored suits seemed always to have just the "right degree of rumple."[48]

The Hogensons' selection of Strandlund proved wise, for he soon developed an important production innovation for the armor plate program that brought widespread acclaim to the company. Strandlund's "press-quench" method for hardening armor plate cut production time from over five hundred man-hours per ton to only twenty man-hours per ton and reduced processing time for one key operation in the plating process from fourteen hours to just six minutes. The innovation was hailed as one of the most outstanding contributions to war production technology in the Chicago production district, earning him an award for engineering innovation from the *Chicago Tribune* and Chicago Vitreous the Army-Navy "E" award for excellence in war production. In September 1943 the Hogensons promoted him to vice president and general manager. He had become the "third brother."[49]

Since arriving at Chicago Vitreous, Strandlund had been intrigued with the company's prewar interest in developing construction systems from enameled-steel panels. In 1932 the company had established Lusterlite Laboratories, a research facility created to develop commercial processes for manufacturing enameled-steel panels. Lusterlite also functioned as an informal subsidiary charged with promoting the use of enameled-steel panels in the construction industry.[50] Industrial applications for porcelain-enameled steel remained limited until the early-twentieth century due to a lack of technical expertise and an inability to produce a quality product on a commercial scale. Beginning in the 1920s, however, several American companies developed new production techniques for the basic component of enamel, a gritty substance known as "frit," and launched important research and development efforts to improve the quality of enamels and reduce production costs. Eager to promote the use of enamels, several firms began to explore architectural applications of enameled-steel panels, first for decoration, then later as structural components in commercial buildings.

The manufacture of porcelain-enameled steel involved several processes. The first step, called fritting, required smelting a variety of minerals, including feldspar, borax, quartz, cobalt, nickel, and manganese in a 2,000-degree furnace in much the same manner as common glass. From the smelter, conveyors directed a molten ribbon of glass through water-cooled rollers. The sudden change in temperature shattered the compound into millions of granules. The frit then traveled through dryers and across magnetic separators to remove traces of iron that caused specks and discoloration in the enamel.[51] The final phase of the process included grinding the frit with water and clay to produce a porcelain enamel solution. The creamy liquid was then ready for application onto steel by dipping or spraying. Only special grades of thoroughly cleaned steel provided an acceptable surface for enameling. After applying the initial coat of liquid enamel, the metal passed first through a drier and then to a 1,600-degree furnace that permanently fused the enamel onto the metal. The resulting product was virtually indestructible, easily maintained, and potentially well suited for a variety of architectural applications.[52]

The two most prominent companies in the field, Ferro Corporation of Cleveland, Ohio, and Chicago Vitreous, formed the core of the U.S. enameling industry in the early-twentieth century.[53] William Hogenson, president and eldest son of Chicago Vitreous founder Emanuel Hogenson, was the driving force behind the efforts to expand markets for enameled steel. A member of the nation's first generation of professionally trained ceramic engineers, he assembled a team of engineer-salesmen at Lusterlite and created the foremost research and development facility in the industry.[54] Hogenson also founded the Porcelain Enamel Institute, the industry's first trade association, and with his brother, Emanuel, Jr. founded and published *Better Enameling,* a trade journal that promoted the enameling industry in general and provided a showcase for the company's line of enameled products for architectural applications.[55]

Like other companies that developed new building components or technologies, the Hogensons realized that Chicago Vitreous's success rested on its ability to educate and evangelize its products to architectural and construction firms. The company would need to demonstrate not only the aesthetic appeal of porcelain-enameled building components but relate that appeal to practical commercial goals such as attracting more customers or decreasing maintenance costs. *Better Enameling* featured numerous, lavishly illustrated articles exploring these themes, including several speculating on the development, and welcome inevitability, of mass-produced porcelain-enameled steel buildings and houses.[56]

Tireless promoters, the Hogensons conceived a dual purpose for Lusterlite Laboratories. In addition to its technical mission, the lab facility itself also served as a promotional vehicle to encourage the use of enamel products in architecture and design. With 10,000 square feet of laboratories and 12,000 square feet of decorative porcelain enamel, it was unparalleled in the industry.[57] While the technical capabilities of Lusterlite drew much attention from the ceramic industry, the striking porcelain-enameled art deco interior drew rave reviews from the design and architectural trade press.[58] Each room featured a different decorative scheme that emphasized eye appeal as well as functionality. Combining form with function throughout the building, the company created an attractive work environment while demonstrating architectural applications of its products.[59]

Lusterlite Laboratories perfectly reflected the visual aesthetic of the era. Streamlined, art deco elegance pervaded industrial design, urging consumption on the depression-ravaged American consumer by appealing to high style and by employing design cues borrowed from airplanes and other "high-tech" harbingers of an exciting future. Even the most mundane objects were streamlined and stylized by such notable industrial designers as Raymond Loewy, Walter Dorwin Teague, and Norman Bel Geddes, often with great commercial success. The home appliance industry felt the full force of the streamlining movement, and as a major supplier of frit to companies manufacturing porcelain-enameled washers, dryers, and refrigerators, Chicago Vitreous was well attuned to the commercial potential of the streamlined design aesthetic. That the built environment should escape the trend—a trend that spoke to the hope for a more prosperous, and indeed more organized future—seemed unthinkable.[60]

The notoriety of Lusterlite Laboratories, coupled with an aggressive sales ethic, enabled Chicago Vitreous to establish important ties with architectural firms. Despite depression conditions, the fledgling business produced a small profit.[61] In addition, Chicago Vitreous became widely known as an innovator in the field, and by the late 1930s many companies sought its services in a variety of construction projects. Architectural enamel came into widespread use in facades for commercial buildings, stores, theaters, diners, and gasoline service stations.[62] The company's foray into porcelain-enameled architectural elements established both its manufacturing competency and its reputation within the industry, as well as provided a commercial context for more ambitious ideas to come.

The relationship between Chicago Vitreous and the oil industry proved especially important, for it signaled the beginning of the company's involvement in commercial structures—in this case filling stations—made entirely of porcelain-enameled steel panels hung on a steel framing structure.[63] An important technological leap beyond products used only for facades, manufacturing filling stations forced the company to design a unified configuration of enameled steel and, more important, to devise schemes for efficient, uniform production.

Between 1938 and 1940 Chicago Vitreous contracted with several major oil companies, including Standard Oil of Indiana, Shell, and Sacony Vacuum, to produce porcelain-enameled steel service stations in a variety of designs. Porcelain-enameled steel, with its shiny, crisp look and minimal maintenance cost, met both stylistic and economic requirements. To promote brand identification among their customers, each oil company desired a distinctive, yet uniform style for their stations.[64] In 1937, Chicago Vitreous organized a subsidiary, the Porcelain Products Company, to handle the financial and organizational demands of the new endeavor.[65]

The creation of the subsidiary also formalized the firm's forward integration into the construction business. Forward integration is a commonly used business tactic to expand market applications for a core product while retaining control over raw material costs. Other firms in the enameling industry pursued similar tactics. In partnership with ARMCO Steel, Ferro Corporation had sponsored the development of a model porcelain-enameled steel house for the 1933 Century of Progress Exhibition in Chicago. As previously noted, a number of metal firms experimented with housing designs with varying degrees of interest and success in hopes of developing new markets to offset their depressed core businesses and absorb excess productive capacity.[66]

With Lusterlite Laboratories providing a technical base, the Porcelain Products subsidiary quickly introduced several important innovations that reduced the cost and improved the performance and attractiveness of enameled-steel buildings. First, company engineers developed new enameling compounds that could be cured, or adhered, to steel at lower temperatures. This enabled firing at temperatures below the critical point at which steel would bend or distort. Additionally, the new compounds proved stronger and more elastic than previous enamels and could be sprayed onto the steel in thinner coats, significantly reducing production costs.[67] The company also developed high-speed production machinery and set up the first assembly line for enameled-steel buildings in the United States. Continual processing through the use of mechanized conveyors, forming presses, automatic furnaces, and specialized welding machinery provided production volumes necessary to realize economies of scale.[68]

Despite significant progress toward the mass production of enameled-steel buildings, the war delayed development of these technologies. Strandlund now saw an opportunity. Looking to ease the company's reconversion and tap into postwar opportunities in the commercial construction industry, he focused energies on renewing the division's prospects. He surveyed the industry, assessed future trends, and eventually presented a vision of the panels not merely as architectural adornments, but as a structural system unto themselves. The implications were clear. Commercial buildings could be manufactured by assembling units of prefabricated porcelain-enameled steel panels. A virtually indestructible, maintenance free, and potentially elegant structure was possible. Further, by bringing more of the construction process under the control of the factory, rationalization and standardization benefits abounded. If the process could be successfully scaled up, it could have profound commercial implications. Strandlund wasted few opportunities to inform the industry of his hopes for the future of enameled-steel panel construction.

In 1945 he also patented a new method for joining enameled-steel panels. This innovation, based on interlocking U-shaped recesses at the ends of the panels, produced a tighter seal than any previous method and significantly reduced the time needed for assembly at the job site. This was the first of many technical innovations in prefabricated building technology that Strandlund devised and one that provided an important basis for commercial production of the Lustron house.[69] At war's end the process of reconversion to peacetime manufacture presented daunting challenges to Chicago Vitreous. Fortunately, the company had secured a contract to produce a new line of filling stations for Standard Oil of Indiana, which provided substantial up-front payments to aid Chicago Vitreous's reconversion process. However, steel remained a controlled commodity under the authority of the Civilian Production Administration (CPA), the peacetime successor to the War Production Board.[70]

Government Entrepreneuring

The task of petitioning the CPA for a steel allocation fell to Strandlund. Arriving in Washington in June 1946, Strandlund's initial confidence in his ability to win the day quickly evaporated as CPA officials all but laughed in his face when he asked for a steel allocation for commercial construction. An admitted political neophyte, Strandlund confessed ignorance of potential bureaucratic roadblocks. He reported to the Hogensons that "the odds of obtaining any steel for our operations seem insurmountable."[71] Yet Mr. Strandlund had arrived in Washington at an opportune moment with the Truman administration in the midst of the veterans' housing crisis—a crusade in which Strandlund was about to play a prominent role.

During one of his meetings at the Civilian Production Administration, Strandlund mentioned that Chicago Vitreous could produce prefabricated houses with the same technologies used for commercial buildings. Intrigued, CPA officials pressed him for details and suggested that he contact the NHA for assistance in formulating a business plan. Wyatt heard about Strandlund from R. Harold Denton, head of the Materials and Equipment Development Division of the NHA and personal friend of the Housing

Expediter.[72] Denton, a Yale-trained lawyer and an expert on housing issues, would play an important role in the Lustron saga. One of many talented young attorneys who hitched their careers to the New Deal, Denton had thought broadly and deeply about industrialized housing. His experiences at the Tennessee Valley Authority, Department of Labor, Home Loan Bank Board, and the NHA led him to embrace the belief that factory-produced housing could modernize the hidebound and inefficient residential construction industry. Denton hoped that mass production could both create and serve a new market for well-made, affordable single-family houses—just the kind of houses that the average returning veteran longed to find. In his view, the field was wide open and the time ripe for the right man with the right idea.[73]

Strandlund and Denton met in early August to outline broad strokes. They had much in common, and Denton was immediately impressed with Strandlund. Strandlund held sway over Denton's staff as well and spent hours discussing various aspects and implications of the proposed venture. Denton became Lustron's most ardent advocate within the NHA, often acting as point man for the company in the bureaucratic struggles to come. Confessing his admiration for Strandlund's idea, he noted that elements of the house could also be used in conventional housing, an application that potentially broadened the scope of the enterprise. Prospective economies of scale also attracted Denton's interest, as well as the production flexibility inherent in the use of enameled steel, a process easily adapted to large-scale manufacture. Strandlund's enthusiasm and energy also impressed Denton, and he told his superiors at the NHA that he believed Chicago Vitreous could begin production as early as January 1947, pending quick approval of a market guarantee and an RFC loan. Denton also promised Strandlund that the NHA would assist in obtaining a suitable production facility and mentioned the possibility of acquiring a large plant in the Chicago area from the War Assets Administration (WAA).[74]

After the meeting Strandlund returned to his Mayflower Hotel suite and spent a sleepless night formulating preliminary production plans. If federal funds could be secured and market guarantees obtained, the government would assume most of the initial risk and he would deliver a revolutionary product. It was a winning proposition for every party—himself, his company, and the country. Perhaps his thoughts turned to the prospect of his contributions to the struggle for veterans' housing and the praise it would engender from a grateful nation.[75] Returning to Chicago the next day, Strandlund conferred with the Hogensons, who agreed to provide $100,000 for the design of a prototype house and the development of detailed engineering, manufacturing, and marketing plans. He also suggested that they rename the Porcelain Products Company "Lustron," a variation of Lusterlite and a contraction of the procedure of putting "luster on" steel through the enameling process. Chicago Vitreous had used the Lustron name occasionally since 1936 and registered the trademark in October 1937. The Lustron house owed its primary technological debt to the enameling industry, and its development evidenced an integration of advanced enameling technology, steel fabrication, and innovative architectural design.[76]

With initial plans laid, Strandlund commissioned two young architects, Morris H. Beckman and Roy Burton Blass, to design a prototype house. Both had worked with Chicago Vitreous in the design of commercial buildings and were familiar with the firm's design and production capabilities. Beckman, a 1938 graduate of the Massachusetts Institute of Technology, had achieved quick success in his field, rising to the position of chief draftsman at the internationally renowned architectural firm of Skidmore, Owings and Merrill. He had left Skidmore in 1945, and with Blass as his partner, established the firm of Beckman and Blass in the Chicago suburb of Wilmette.[77]

Blass and Beckman, intrigued with the opportunity to influence postwar housing design, sought to provide a "revolutionary" vision for the postwar family. One of the strongest consumer desires reflected in contemporary housing surveys was for flexible interior designs with open spaces and multipurpose rooms. This signaled a new interpretation of domestic space, a consequence of such postwar expectations as larger, child-centered nuclear families, suburban living, and increased consumption. The architects envisioned that the Lustron house would both provide and reflect a "total approach to living" by accentuating new "domestic technologies" within an open, rambling plan that emphasized family activities.[78]

Blass and Beckman chose a ranch-style house design for a number of reasons. Long associated with cost-effective construction, the ranch form was similar to the type of commercial building that Chicago Vitreous knew how to design and produce in a cost-effective manner. The simplicity of design and adaptability to a wide variety of geographic locations and lot characteristics gave the Lustron house a broad potential market as well as a visual familiarity that the architects hoped would mitigate potential negative criticism of its steel construction. The inherent efficiency of the ranch style represented modularity and modernity, a reduction of housing to its most essential features without sacrificing livability. It looked good in many orientations and could accommodate a range of interior plans in line with emerging consumer tastes.[79]

A modern house for modern American families, the ranch had gained popularity among architects since the early-twentieth century due in part to the work of Irving Gill and Frank Lloyd Wright. By the 1940s the term *ranch house* was familiar throughout the profession, and young architects like Blass and Beckman were intrigued with the prospect of using the ranch design to synthesize the latest in construction efficiency while meeting the needs of modern family life. For Lustron, the association of the ranch with western and especially West-Coast lifestyles evoked freedom, individualism, and an easygoing pioneerism that gave potential buyers a favorable

A woman relaxes in the front yard of a Lustron demonstration house.
Photo credit: Ohio Historical Society

context in which to evaluate the house and a chance to deemphasize the standardization inherent in prefabricated construction.[80]

The prototype design, called the "Lustron Esquire," was a 1,025-square-foot, one-story, two-bedroom, gabled-roof ranch with a bay window and recessed side porch. It incorporated numerous innovations designed to appeal to postwar expectations for comfort and convenience. A key element in unifying function and utility with the "open space" concept was the extensive use of built-in storage, including kitchen cabinets, bookshelves, a china closet with a pass-through feature from the kitchen to the dining alcove, a vanity in the master bedroom, and large closets throughout. In fact, the architects devoted over 20 percent of total wall space to built-ins.[81] Blass and Beckman also included a utility room adjacent to the kitchen containing the furnace, water heater, and space for a freezer. The Esquire's all-electric kitchen included a combination dishwasher/washing machine dubbed the "Automagic." Manufactured by the Thor Corporation, the Automagic was a true novelty and excellent selling point since few homes in 1946 had dishwashers.[82]

Structurally, the house consisted of exterior and interior enameled-steel panels bolted to a steel skeleton and fixed to a concrete slab foundation. Wall framing was accomplished by joining exterior and interior metal studs with spacer plates, which provided a thermal break for insulation. Diagonal support braces tied the system together.

The exterior contained twenty wall framing sections and ten corner assemblies. Ten trusses placed at four-foot intervals formed the roof structure.[83] The exterior panels measured two feet square, while interior panels measured two feet wide and ran from the floor to the ceiling, except for two-foot-square panels in the bathroom and kitchen. Interestingly, interior panels were grooved to resemble wood paneling. Four-foot-square ceiling panels completed the interior. Porcelain-enameled window frames, doorjambs, gable ends (also grooved to simulate wood), gutters, soffits, and roofing tiles assured that the house would remain virtually maintenance free. The dominant feature at the front of the house was a three-piece bay window, and decorative trellises supported the corner of the recessed porch.[84]

The Lustron prototype successfully addressed a significant shortcoming in earlier steel designs with several energy-saving features. Each exterior panel was backed by 1.5 inches of fiberglass insulation. Coupled with the thermal break between the exterior and interior walls, this design ensured sufficient heat retention to bring heating costs within conventional norms. Additionally, by attaching insulation board to the underside of the roof trusses, a plenum chamber was created between the insulation boards and the ceiling panels. An electric oil-fired forced-air furnace used the plenum chamber to create a radiant heat source for the entire house.[85]

Displaying these innovations to best effect was an important concern, and Strandlund arranged to assemble the house on the grounds of the nearby Hinsdale Nursery in the midst of a formal garden. Careful attention to the relationship between industrialized housing and the natural environment had been a missing component in many prefabricated housing ventures. That Strandlund appreciated the connection between the two evidenced a broader view of the enterprise, though he could not have known just how crucial the issue of "proper" placement of these unusual houses would become in the future.

While the architects and Chicago Vitreous engineers reviewed the prototype and formulated production plans, Strandlund returned to Washington, his briefcase bulging with blueprints, production estimates, marketing strategies, and financial requirements for full-scale production of the Lustron house. Confirmed meetings with the CPA, Housing Expediter Wyatt, and additional officials from the NHA demonstrated serious governmental interest. Strandlund was encouraged by the prospect of federal support for Lustron and confident in his abilities to close the deal.

Lustron submitted an application to the NHA on September 23, 1946, requesting a guarantee for 30,000 houses at an F.O.B. price of $4,341 each. The price reflected the following components: $372 for labor; $2,454 for materials; $1,099 for overhead; and $454 for pretax profits. Strandlund promised to produce at least 30,000 houses by December 1947 and estimated a production rate of 400 houses per day beginning in January 1948. The figure of 400 houses per day represented the estimated break-even point for Lustron.[86] That the company could produce complete housing units as well as materials and components for other housing producers, both prefabricated and conventional, was especially attractive to the NHA. As Denton had noted, this held great potential for broad marketing.[87]

NHA officials also noted that Lustron was prepared to begin pilot production immediately since Chicago Vitreous's commercial building plant stood effectively idled by the lack of a steel allocation. Strandlund estimated that within a month the company could produce ten houses per day at the Cicero plant, and reported that Lustron had already arranged preliminary contracts to supply other prefab companies with porcelain-enameled roofing tiles and discussed the possibility of supplying bathroom units as well. Strandlund had also contacted several conventional builders who expressed interest in using Lustron's steel framing system.[88]

The next challenge facing Lustron was the acquisition of a production facility capable of supporting large-scale manufacture. Chicago Vitreous engineers estimated that the firm would need at least 2.5 million square feet of plant space to achieve the necessary economies of scale. Both Strandlund and the NHA believed that the six-million-square-foot Dodge-Chrysler plant in Chicago presented the best possibilities. Located only four miles from Chicago Vitreous, its proximity meant that engineers and technicians could supervise Lustron's operations during the crucial startup phase. Further, since Chicago Vitreous planned to supply the majority of Lustron's frit requirements, shipping costs of the bulky material would be significantly reduced. An important center of the enameling industry, the Chicago area also promised an ample supply of experienced labor. Strandlund estimated that the venture would require at least 20,000 employees for full production. Ready access to raw materials, as well as ease of distribution, would also be guaranteed by the location's existing rail facilities. Lustron's manufacturing operations required unobstructed floor space for the assembly line, high ceilings to accommodate huge enameling furnaces, and fireproof construction. The Dodge-Chrysler plant met each of these criteria.[89]

During the war the Dodge-Chrysler plant had produced B-29 aircraft engines with the aid of approximately $171 million in federal subsidies. Declared a surplus facility at the end of the war, the War Assets Administration sought companies that would use it "to the best interests of the Government and the American people."[90] What could be more important than housing? The answer, in true American fashion, appeared to be automobiles. Another capitalist with big ideas, dreams, and ambitions—Preston Tucker—also had designs on the facility. Throughout the summer of 1946, the RFC and the WAA had been negotiating with Tucker, who wanted the Dodge-Chrysler plant to manufacture the innovative Tucker Torpedo. NHA officials and Housing Expediter Wyatt remained optimistic about acquiring the plant because they believed that the RFC and the WAA would consider housing a greater national priority than automobiles and because it seemed unlikely that Tucker could raise over $15 million in private capital to fulfill his responsibility under a preliminary lease agreement signed in July.[91]

Yet this was far from ideal, since it limited space for future expansion. As a hedge, the NHA and Strandlund decided to earmark surplus war plants used by aircraft manufacturer Curtiss-Wright in Cincinnati and Columbus, Ohio, two of fifty surplus plants promised by the WAA for conversion to housing manufacture. At five million square feet and three million square feet respectively, the plants offered sufficient capacity and access to equally sophisticated labor markets.[92]

Obtaining a suitable plant represented only a part of the startup process. Equipping it with the necessary machinery presented equally formidable obstacles. Strandlund estimated a cost of approximately $22 million and requested that the Defense Plant Corporation, a subsidiary of the Office of War Mobilization and Reconversion, purchase the machinery and lease it to Lustron with an option to buy. This was an extension of the typical agreement used by the government during the war to encourage conversion to defense manufacture. However, such an arrangement would require the approval of the RFC as guarantor of the loan. Denton urged the NHA to assist Lustron in securing RFC approval.[93] The matter of preproduction expenses also caused concern. Estimates for engineering, construction, and organization of the plant totaled approximately $10 million for the period from November 1946 to July 1947. Lustron hoped for a series of advance payments from the RFC, though NHA officials realized the prospects for quick action from the RFC were dim. Private financing would have to carry the load until government funding could be secured.[94]

Indeed, by September 1946, Chicago Vitreous had already invested over $500,000 in the development of the prototype, including preparation of dies and equipment for the production of ten houses per day. Strandlund, Emanuel Hogenson, and William Hogenson had also invested approximately $100,000 of their personal funds in Lustron, of which $36,000 remained in the account. The $36,000 figure is significant, for as the venture made its way through the bureaucratic maze, opponents seized on the number, claiming it represented the total equity value of corporate and personal investment in Lustron.[95]

With production and financial plans progressing, the approval of a market guarantee represented a vital first step in securing the government's participation in the mass production of Lustron houses. The favorable assessment of James L. Pease, director of the Prefabrication Production Branch of the NHA, was crucial. Reviewing all aspects of Lustron's business plan, Pease concluded in mid-October 1946 that the firm had "demonstrated its ability to produce houses at the proposed rates."[96] The company's intention to manufacture or furnish all components of the house except the glass for windows and the stove and refrigerator especially impressed Pease. He also noted that Chicago Vitreous's ongoing research and production expertise had resulted in lowering the price of enameled-steel construction to the point of commercial viability in the housing market. Lustron, Pease asserted, represented the "best bet" in the field.[97]

However, the company's marketing plan raised serious questions and became one of the greatest unknowns of the entire venture. Strandlund's initial marketing program included a national distribution network, sales promotion campaigns, and a system for closing sales and monitoring the erection of houses on the local level. The company would assign an exclusive dealer to each city with a population over 50,000, assuring at least 600 dealers to start. Using the automotive industry as a model, Lustron dealers would be organized into ten regional offices intended to enforce administrative control from the central office. A promotional budget of at least $200,000 would be used for an advertising campaign focusing on such national weekly magazines as *Life, Look, Better Homes and Gardens,* and *Collier's.* Lustron would not have to start from scratch, since key personnel from Chicago Vitreous could guide the formation of Lustron's sales and distribution systems in the initial stages.[98]

The relationship between the company and its dealers was intended to provide mutual financial and operating advantages, with the ultimate goal of introducing the Lustron house into conventional residential areas. Here Lustron would face numerous obstacles. Key factors included the lack of experience in the field of domestic real estate, the sheer novelty of a house made from porcelain-enameled steel, and an initial lack of understanding of the complex financial arrangements inherent in marketing prefabricated houses.[99]

Strandlund intended to reinforce the dealer relationship by training "house erection teams" that would remain on call to assist dealers and customers during the assembly phase. Additionally, he contacted a number of large-scale suburban developers across the nation to arrange volume discounts. This strategy was intended to serve the dual purpose of guaranteeing the sale of Lustron's initial output, as well as to ease the transition of the house into conventional and suitably zoned suburban settings. As Pease observed, the firm seemed able to "nurse along the new venture during the critical first months when this new product will have to prove its worth."[100]

Ominously, though, Pease cautioned that the unprecedented volume of production would have to be matched by an equally unprecedented commitment to service on the local level. Other potential problems at this level included the attitude of local labor organizations, contractors, and building code compliance officials. Nevertheless, Pease recommended that the NHA enter into negotiations with Lustron, but only for an initial market guarantee of 15,000 units, half the original request. Although disappointed, Strandlund understood the bureaucratic prudence and remained confident that the various roadblocks could eventually be resolved. After all, during the war, he had brought Chicago Vitreous's tank turret business up to full production from scratch in less than three months.[101]

Allies, Enemies, and a New Start

Indeed, by October 1946, Lustron's future appeared bright. Press reports about its plans elicited favorable responses, especially from veterans' organizations. Wilson Wyatt, fearing a loss of momentum in the VEHP, had consistently urged veterans to use their political influence to encourage government aid for the immediate production of prefabricated houses. With Lustron facing a major hurdle in securing RFC financing, the Housing Expediter issued a rallying cry. Following Wyatt's lead, Omar Ketchum, director of the National Legislative Service of the Veterans of Foreign Wars, appealed to the RFC: "Veterans of the nation have waited for many months for a solution to the acute housing shortage. We believe that the only solution is through mass production." Ketchum urged the RFC to provide 100-percent-guaranteed loans for prefabricated housing as they did for defense production during the war. "If we can start companies, such as the Lustron Corporation of Chicago . . . the problem of price will be solved and the VEHP saved," he wrote.[102]

Buoyed by the support of the VFW, on October 22 Wyatt submitted an application to the RFC for a $52 million loan for Lustron, a step that initiated a serious political struggle between Wyatt and the RFC over Lustron's future. The struggle also reflected a larger controversy over the direction, and indeed continued existence, of the VEHP. Wyatt had essentially bet his job on the quick development of prefabricated housing, and he was prepared to use all of his official powers and political skill to convince the RFC to finance this strategy.[103]

George Allen, a director of the RFC and close friend of President Truman, handled Wyatt's request. A fiscal conservative and influential member of the board, Allen advanced a number of ideological and political objections that resulted in the rejection of the application on October 31. The RFC's opposition centered on the "appalling" ratio of invested capital in Lustron in relation to the $52 million request. The RFC based its analysis on an equity figure of $36,000, the amount remaining in the Strandlund/Hogenson personal investment account, and discounted or ignored nearly $500,000 of total investment capital provided by Chicago Vitreous and the three men. Naturally, any prudent loan officer would agree that an equity ratio representing 7/100ths of 1 percent of total debt should merit rejection. As RFC Chairman Charles B. Henderson observed, "if the enterprise were unsuccessful, substantially all of the risk and loss would fall on the Government whereas if the enterprise were successful the profits to the owners would be enormous and out of all relationship to their investment or risk." If Lustron met its production estimates for 1947, Henderson calculated that Strandlund and the Hogensons could reap a 14,000 percent return on investment.[104]

Without a profit-sharing arrangement, Henderson argued, the loan was actually a grant or a subsidy. The RFC simply could not legally offer grants to industry, nor would it welcome the inevitable scrutiny such an arrangement would elicit. Clearly, the equity situation offended the political sensibilities of a majority of its directors. Further, the RFC interpreted the powers of the VEHP narrowly, insisting that the only authorized measures were premium payments to stimulate the production of building materials in short supply and guaranteed market contracts for prefabricated houses. This implied that Wyatt could ask, but could not force, the RFC to provide loans for prefabricated housing ventures. Henderson attempted conciliation by assuring Wyatt that the rejection "in no way reflects any lack of understanding or appreciation of the importance of the VEHP, nor does it preclude consideration of subsequent applications from Lustron or others."[105]

Wyatt, frustrated by the RFC's action, responded on November 1 in a letter to Henderson that emphasized the need for immediate action to ease the housing crisis. "We do not feel that the failure of any company to meet any particular ratio of equity capital should be a basis for declining a loan application for a sound house or new type of material that is needed as part of the VEHP," Wyatt wrote. Challenging the RFC's low estimate of Lustron's equity capital, he asserted that the firm actually possessed approximately $5.7 million in equity contributions, including $500,000 in prototype development costs; $5 million for patent rights; $200,000 for machinery, tools, and dies; as well as the $36,000 in the Strandlund/Hogenson private investment account.[106]

Wyatt then presented several alternatives to encourage the RFC to reconsider the loan. First, he proposed a joint financing arrangement by which a group of Chicago banks would provide $20 million, an arrangement that would raise Lustron's equity contribution to $25.7 million against the $52 million request.[107] Next, Wyatt reminded Henderson of his power to impose "appropriate conditions" on any loan agreement to assure that the purposes of the VEHP would be "fully accomplished." In the case of a firm with little or no equity, such conditions included limiting the amount of profit earned or dividends paid in order to accelerate repayment of the loan; limiting executive salaries; or limiting depreciation allowances for capital equipment. In fact, Wyatt claimed, Carl Strandlund had already agreed that Lustron would not keep any profits or advance any dividends to investors until the entire amount of the RFC debt was retired. Certainly this would assuage any fears regarding Lustron's potential for quick profit at government expense. In a conciliatory yet firm tone, Wyatt cautioned the RFC Chairman against obstructing the VEHP: "We feel that [these conditions] meet the objection raised and make possible the accomplishment of our paramount objective which is to enable [Lustron] to get into production and make its vitally needed contribution toward the alleviation of the housing emergency."[108]

The Housing Expediter also enlisted his allies at the VFW in the campaign against the RFC. On November 13, 1946, representatives of the organization presented a letter from VFW Commander-in-Chief Louis E. Starr to President Truman and issued a press release calling on the president to "use his full authority to assure Wilson W. Wyatt the cooperation of every governmental agency."[109] In his appeal to Truman, Starr bemoaned the apparent lack of support for the VEHP "promised by government agencies who were supposed to be our allies." Specifically blaming the RFC, he noted that Wyatt's commitment to prefabrication was the only hope to provide veterans with affordable housing in the least amount of time possible.[110]

Lustron soon had another resource to leverage in its quest for federal support. By mid-November, the prototype house was complete and ready for public inspection. This proved extremely important, since Strandlund could now invite interest groups, key politicians, and the general public to "see for themselves" the promise of Lustron. He carefully orchestrated Wyatt's visit, and Wyatt left Chicago "very impressed," noting that "it measured beyond our expectations." The guest book soon bulged with glowing comments and purchase offers. The house also received favorable reviews from the local and national press, with several articles commenting that the promise of prefabrication had finally been fulfilled.[111]

Strandlund made sure that major veterans' groups received special tours. Representatives from the VFW, the American Legion, and AMVETS soon offered their endorsement, stating that "it met the veterans' needs ideally." In an astute public relations move, Strandlund offered to sell the house to Elmer R. Theidel, a recently married Navy veteran. Waving happily to America from the porch of the "Esquire," the Theidels represented the very definition of the postwar domestic ideal—an image the national press devoured.[112]

Concurrent with the struggle to obtain RFC financing, Wyatt and Strandlund faced a flare-up in their battle with the WAA and Preston Tucker over the acquisition of the Dodge-Chrysler plant for Lustron. Both Wyatt and Strandlund believed that the company could begin production more rapidly at this plant than at any other. Time was a crucial consideration for Wyatt. Consequently, he resolved "to go to the mat" to secure the facility, using his powers under the Veterans' Emergency Housing Act freely if necessary.[113]

In Wyatt's opinion, awarding one of the largest plants in the world to Tucker was absurd. Raw material and parts shortages, not plant capacity, represented the principal factors limiting the production of automobiles. Nevertheless, Tucker and the WAA had forged a good working relationship, and WAA officials soon adopted a defensive stance against the Housing Expediter. Each party also tapped into the Washington grapevine as much as possible for information about the potential responses of the RFC, which held the future of both enterprises in its control.

Negotiations between the Office of the Housing Expediter and the WAA began on September 11, 1946. On that date Strandlund contacted Wyatt and formally requested that the plant be "frozen" for Lustron's use. He indicated that the *Chicago Tribune* had reported the previous day that the WAA had given the Tucker Corporation ten days to meet the $15 million private capital requirement under their preliminary lease agreement signed in July. Wyatt immediately contacted the WAA regarding the status of the plant, and it informed him that "Tucker had the plant but was experiencing difficulties with financing and that there was always a remote possibility that the deal would fall through." In view of the preliminary stages of the NHA's negotiations with Lustron and because the WAA seemed committed to Tucker, Wyatt did not invoke his power to freeze the plant.[114]

With the plant controversy at a standstill, an embattled Wyatt sought public forums to present his point of view on the RFC loan for Lustron and the future of the VEHP. The ideological and political climate of the nation undermined his efforts. A cornerstone of the VEHP was the continuation of price controls and production and allocation restrictions on building materials. Yet the American public longed for an end to wartime controls and a return to normal economic life. The electorate expressed such desires on Election Day, giving the Republicans control of both houses of Congress.

Responding to the election results, facing pressures from fiscal conservatives within the administration, and fulfilling his promise to end controls as soon as possible, Truman abolished most price controls, including those on construction materials, on November 9. Unless he also lifted controls on the price of new homes quickly thereafter, construction materials would inevitably flow into commercial markets.[115] Wyatt of course realized that Truman's actions effectively curtailed the VEHP. The absence of price controls, Wyatt lamented, "would mean inevitably that the Veterans' Emergency Housing Program, as now planned, would be unattainable." Truman's message ending controls also contained an ominous note: "The removal of price ceilings on building materials will obviously necessitate a change in approach to some of the problems in the housing program." The Housing Expediter recognized that he probably was not part of the "change in approach" to which Truman referred.[116]

Wyatt immediately responded to the assault on the VEHP. He appealed to the nation in a radio speech over the Mutual Network on November 14, warning of the "grave danger" of removing controls and predicting that few veterans would be buying houses in 1947. The public responded indifferently to his message, which disappointed Wyatt. Yet he understood their skepticism about further appeals when all they wanted was affordable housing as soon as possible.[117] Nevertheless, Wyatt pressed forward as he testified before a Senate subcommittee investigating national defense issues on November 25. He presented an eloquent defense of the VEHP, leveled harsh criticism at the recalcitrant RFC, and issued a plea for congressional support of the prefabricated housing industry. "These are not ordinary times," Wyatt entreated, "the need of veterans for homes is not an ordinary need; it is a desperate need. To enable private enterprise to meet this need, the government must assume the financial risk. Banking as usual begets business as usual. Neither is enough to meet the housing emergency."[118]

Wyatt also complained that the controversy over the Lustron loan was clouding the larger issue of support for other types of prefabricated construction, such as stressed-skin plywood panels, metal-covered plywood, concrete, and aluminum. The RFC's stance, he added, had also impacted manufacturers of both new and conventional types of building materials, such as gypsum board and substitutes for hardwood flooring. "Conventional homebuilders are being hurt by the same inaction that is delaying the production of industrialized homes," Wyatt argued.[119]

Finally, on November 27, Wyatt presented his VEHP forecast for 1947 to President Truman, along with a statement threatening resignation if the president did not accept its terms. Wyatt called both for the continuation of the price ceiling on new houses and for up to 100-percent-guaranteed government financing for prefabricators—politically unrealistic requests that reflected Wyatt's despair. During the last week of November, Truman summoned Wyatt and George Allen to the White House several times in an attempt to resolve the matter. "We are in complete disagreement," Wyatt and Allen told the press. Publicly, Truman remained significantly silent and left Washington for a short vacation.[120]

Wyatt resigned on December 3, offering a tersely worded letter that failed to thank the president for his assistance. The following day Truman accepted Wyatt's resignation and expressed gratitude for his "outstanding" efforts on behalf of the nation's veterans. The VEHP was dead, and within a week Truman removed the $10,000 ceiling on new house prices and announced a relaxation of rent controls. Clearly, Truman believed that the VEHP was no longer compatible with an expanding peacetime economy or with the public's desire to end all wartime controls.[121]

Wyatt's resignation placed the Lustron experiment in serious jeopardy. Strandlund issued a press release on December 6 reiterating his commitment to the project and maintaining the necessity of some form of government financing to help Lustron begin production. However, Strandlund also offered a key compromise. He announced that Lustron would be willing "to operate on a scaled down basis," which would "reduce our requirement for assistance accordingly." Here was the genesis of the strategy that ultimately influenced the RFC to reconsider financing the venture. "We are anxious to cooperate on any basis which will make possible the production of Lustron homes," Strandlund asserted.[122] Additionally, he offered to drop the request for the Dodge-Chrysler plant in favor of obtaining a smaller facility, such as one of the Curtiss-Wright plants in Cincinnati or Columbus. Abandoning the original production estimate of 400 houses per day, Strandlund announced that "the new scaled down Lustron program" would initially produce 100 houses per day and require approximately $16 million in government financing rather than the original request for $52 million.[123]

The potential for reconsideration by the RFC greatly increased with the resignation of George Allen from its board in mid-December. Exploiting this opportunity, newly appointed Housing Expediter Frank R. Creedon aided Strandlund in the preparation of an application to the RFC for a $15.5 million loan and indicated that the NHA would support a market guarantee for 15,000 homes during 1947. RFC chairman John D. Goodloe received the application on January 17, 1947, and prudently called for a full audit of Lustron's plans by the New York consulting engineering firm Stone & Webster. Goodloe evidently did not expect a favorable report, but Stone & Webster conferred its blessing on the enterprise, praising Lustron's technical resources and manufacturing expertise.[124]

Proceeding cautiously, Goodloe insisted on attaching several conditions to the loan agreement. Before receiving any federal money, Lustron had to raise $3.5 million in private capital. The RFC also sought the personal guarantee of the Hogensons and Strandlund and a commitment from Chicago Vitreous to support Lustron during the startup phase. With all parties agreeable to these terms, Strandlund sought the aid of the Chicago brokerage house of Hornblower and Weeks to arrange a stock offering.[125]

However, negative publicity surrounding Lustron's original loan request, the resignation of Wyatt, and Truman's removal of price controls created a soft market for Lustron stock. Despite Strandlund's contention that "by the third quarter of 1947, Lustron will be shipping 100 houses a day, 2,500 a month, 30,000 a year," investors, especially venture capital firms, remained wary. Despite intense advocacy, by the end of April Hornblower and Weeks had raised only $840,000. The majority of subscribers represented Lustron's potential suppliers, including several steel firms and appliance manufacturers.[126]

At this point Strandlund had only one option—attempt to renegotiate the loan agreement. Given Lustron's rocky history with the RFC, Strandlund would need all of his sales abilities to convince the agency to increase their participation in the loan from roughly 80 percent to nearly 95 percent. The RFC had only agreed to such a high level of participation in loans made to manufacturers of products critical to the war effort. Sensing political disaster, Goodloe bluntly informed Strandlund that unless he could produce "unmistakable evidence" of congressional and presidential support, the loan could not be made.[127]

When Carl Strandlund arrived in Washington during the summer of 1946, he admitted that he did not even know the name of his Chicago alderman, let alone his congressman. He had undoubtedly acquired a good measure of political savvy since, but without powerful advocates on Capitol Hill and in the Truman administration the chances of saving the loan appeared bleak. However, Lustron retained the support of the major veterans' organizations, and their lobbyists offered to guide him through the political jungle once again. Then, as Strandlund was packing his suitcase, perhaps to leave Washington for good, the phone rang. It was Louis Starr, who informed him that the VFW had arranged appointments with a number of influential congressmen. The future of Lustron now rested on Strandlund's ability to win friends and influence people, skills he assuredly possessed.

CHAPTER THREE

From Politics to Production

> We must remember that mass production of housing and especially a fully industrialized brand of it is without historical precedent.
> —Richard Neutra, *Survival Through Design*

By the end of 1946, Carl Strandlund had become a nationally known figure whose plans to mass-produce a prefabricated porcelain-enameled steel house were regularly celebrated in the media as an important part of the government's solution to the postwar housing crisis. Navigating the political minefields in search of government financing for his company consumed Strandlund's energies during the last months of 1946. He made promises regarding scale, scope, and speed of production that favorably influenced politicians and government administrators. Their continued support could open a floodgate of capital and shelter the nascent enterprise; their opposition or reticence could quickly scuttle the company and ruin Strandlund's personal reputation. Although political battles continued throughout its entire history, in 1947 Lustron won a key struggle that enabled it to begin production, launch a nationwide promotional campaign, and create a corporate hierarchy to implement the transition from entrepreneurial dream to commercial reality.

Swirling about the company was the ongoing struggle over the direction of federal housing policy in the postwar era. Supported by the Truman administration and a bipartisan coalition in Congress, a comprehensive housing bill sponsored by Senators Taft, Ellender, and Wagner promised to guarantee decent housing for every American family. The bill and subsequent amendments provided for the expansion of primary and secondary credit programs, easier access to FHA-insured mortgages, housing research, slum clearance, and the construction of half a million public housing units for low-income families over the coming decade. The prospect of an active federal presence in the housing market for the foreseeable future soon drew familiar objections. Buoyed by the 1946 congressional elections that returned control to the Republicans for the first time since 1931, conservatives in government and business rallied once again to the cause of the private housing industry.[1] However, the ongoing veterans' housing crisis kept the production issue in the forefront of public concern and enabled Lustron to tap into public sentiment favoring federal support of the prefabricated housing industry. It made for some strange bedfellows as the prospect of large-scale factory-made housing began to rattle longstanding relationships within the American housing system.

Friends in High Places

RFC chairman John Goodloe informed Strandlund in January 1947 that the agency needed "unmistakable evidence" of high-level support to secure a loan for Lustron. In response, and with the assistance of VFW Commander-in-Chief Louis Starr, Strandlund intensified his efforts to win the support of key congressmen and administration officials. Senator Ralph Flanders (R-VT), a member of the Banking and Currency Committee, the committee charged with oversight of the RFC, welcomed Strandlund into his office on several occasions during the last two weeks of January 1947. A mechanical engineer and former banker, Flanders was impressed with Lustron's technological base and manufacturing plans. He offered to sponsor Strandlund's testimony before the Banking and Currency Committee in an effort to cultivate support for a reconfigured loan. Other legislators who lent a sympathetic ear included Representative Frank L. Sundstrom (R-NJ) and Senator Joseph McCarthy (R-WI), both of whom served on the Banking and Currency Committees of their respective chambers. John R. Steelman, former administrator of the Civilian Production Administration, then serving as an assistant to President Truman, also emerged as an important source of support.[2]

As Strandlund met with key legislators, his associates formulated a marketing strategy to support the firm's political appeals. The strategy soon engendered favorable responses from the national press. Lustron's participation in a March conference sponsored by the National Housing Committee of the VFW provided a good opportunity to inform the public of the company's intentions. The conference agenda included discussions about the future of industrialized housing and focused on strategies for encouraging the acceptance of prefabrication. On the final day of the meeting, Lustron's public relations director, Charles Prins, outlined the firm's marketing strategy and emphasized several themes for maximum political effect.[3]

Prins told the conference that Lustron planned to focus sales in communities with programs giving preference to veterans' housing and announced that the company would favor veterans by offering them "first crack" at dealerships. Prins also presented an intriguing proposition to housing reformers when he revealed that the company planned to design "prefabricated multiple dwellings that will be offered for low-rent, subsidized public housing."[4] The first mention of Lustron's interest in public housing, Prins's comment was highly speculative and politically motivated to further rationalize potential government investment. In fact, while Strandlund and NHA officials had once discussed the issue, neither the company nor the NHA had given it serious consideration.

Another opportunity to present Lustron in a favorable light arose in March. At the urging of Representative Sundstrom, Strandlund agreed to testify before the House Banking and Currency Committee during its deliberations on H.R. 2549, a bill to amend certain provisions of the Veterans' Emergency Housing Act (VEHA). From Lustron's point of view, the most important aspect of the bill was the extension of government aid to prefabricated housing manufacturers beyond the original June 30, 1947 expiration date of the VEHA. As Strandlund observed, an extension to March 31, 1948, would afford the industry "an opportunity to gear up for production" and provide additional time to "sell" its ideas to wary investors. Although the bill gutted the VEHA's price, wage, and rent control provisions, it maintained the government's power to allocate materials and provide market guarantees for housing, provided that it was intended "to assure preference or priority to veterans of World War II or their families."[5]

Strandlund testified before the committee both as head of Lustron and as president of the National Association of Housing Manufacturers (NAHM). Founded in February 1947, NAHM represented approximately eighteen manufacturers with plans to mass-produce "unconventional housing materials and new prefabricated designs." Primarily a political organization, NAHM focused on lobbying Congress and the Truman administration to secure favorable legislation, financial assistance, priority allocation of materials, and mortgage financing for firms on the cutting edge of prefabrication technology. Along with Strandlund—the primary financial and organizational force behind the association—other key members included Abel Wohlstetter, president of General Panel Corporation; Harry Nagin, president of Reliance Homes; and David L. Krooth, former general counsel of the NHA. As one industry observer commented, "if prefabrication is to mean revolution, these are the revolutionaries."[6]

Strandlund began his testimony with a statement delineating the differences between the "new generation" of prefabricated housing manufacturers, who offered innovative designs and "real mass production methods," and companies that "merely transferred the traditional materials and methods from the building lot to a factory" and "continued to do the same old sawing and nailing under a factory roof."[7] Offering the analogy that the mass production of ready-made clothing had not destroyed the business of custom tailors, he sought to allay the fears of conventional builders who believed that the mass production of prefabricated houses would threaten their livelihoods. Rather, factory-made housing would serve an important social purpose by delivering decent houses to wage-earning families. In Strandlund's opinion, tapping into this vast market also guaranteed the long-term viability of the mass production approach.[8]

Again emphasizing the lack of affordable housing for veterans, Strandlund wondered why they should be "victimized" by a housing industry that continued to build costly houses with "techniques and materials that were popular in biblical times." Traditional construction had served "better-than-average" income brackets well but had failed "Mr. and Mrs. Kilroy"—returning veterans and their families who most often fell into the working and white-collar middle classes. Large-scale mass production, Strandlund argued, was the only method to deliver houses at a price that the average veteran could afford. Since the average annual income of veterans in 1946 was approximately $2,500, the "magic number" for the industry was $5,000. "What this country really needs is a good $5,000 house," Strandlund asserted, wryly echoing former U.S. Vice President Thomas Marshall's famous quote about the country needing a good five-cent cigar.[9]

Most of Strandlund's testimony urged the committee to extend the market guarantee provision of the VEHA into 1948. This was crucial, since market guarantees provided essential collateral against RFC loans and private investment. As Lustron's experience evidenced, attracting venture capital from private or public sources had proven extremely difficult even with market guarantees. Without them, it seemed impossible. "No single action that the Congress could take would be of greater benefit to the solution of our housing problems or cost the taxpayer less," Strandlund implored.[10]

The relationship between a viable prefabricated housing industry and the issue of rent control, also under consideration by the committee, provided another compelling argument for government support. The need for rent controls would be eliminated, Strandlund noted, by the availability of affordable mass-produced houses. Additionally, he argued for maintaining limitations on nonresidential construction and recommended the continuation of allocation restrictions on raw materials necessary for innovative approaches to prefabrication.[11]

During questioning, committee members rarely challenged Strandlund and frequently lobbed friendly informational queries seemingly designed to cast Lustron and the prefabricated housing industry in a favorable light. Representative Sundstrom (R-NJ) gushed that he was in "100 percent agreement" with the need for governmental support of the industry. Sundstrom's questioning also reflected his intent to ease fears that market guarantees and RFC loans might be considered unnecessary government subsidies to an untested industry. Strandlund assured the committee that nothing could be further from the truth. If Lustron failed, the government could recoup its investment by terminating the market guarantee and liquidating the company's assets. Market guarantees, he added, served the same purpose as FHA mortgage insurance for conventional builders, which provided access to capital for on-site construction. Since such mortgage insurance was unavailable to factory-made housing, the market guarantee was simply another type of incentive that reflected a continuation, not an aberration, of governmental housing policy. Besides, Strandlund argued, once the industry was adequately capitalized, it would no longer require direct assistance.[12]

Strandlund's testimony attracted the attention of administration officials. White House aide John Steelman and RFC chairman Goodloe discussed the Lustron loan during several meetings in May and June 1947. The company's inability to raise sufficient private capital to meet the terms of the January agreement caused Goodloe to remain wary. Yet Strandlund's largely successful quest to marshal political allies on Capitol Hill, and the favorable press that accompanied his efforts, increased pressure on the RFC to consider reducing Lustron's equity requirement. Although Steelman sympathized with RFC concerns, he revealed strong support for the company, commenting that "the Lustron method represents the fullest application of mass production techniques to the housing problem." Additionally, Steelman observed that even a reconfigured loan agreement would remain in line with congressional desires, expressed in the passage of H.R. 2549 as the Housing and Rent Act of 1947, to employ "unusual measures to further the factory-built housing program."[13]

Goodloe realized that Steelman was expressing the views of President Truman, but he also demanded written confirmation of administration policy regarding the Lustron loan. During the last week of June, as the RFC board again considered the loan, Steelman sent a letter to Goodloe confirming the president's position. The letter outlined the terms of the new agreement: Lustron would provide $500,000 in private capital, and the RFC would increase its participation to $15.5 million. "I have discussed this matter with the President and he has authorized me to state that the views expressed herein meet with his approval," Steelman wrote; "I believe, therefore, under all the circumstances, that the loan should be made."[14]

The RFC board unanimously approved the loan on June 30. "I know of no loan application since I have been connected with RFC that has received more thorough and more careful consideration," Goodloe observed. Although Truman's approval no doubt influenced the board, Goodloe detailed additional elements that led the RFC to the decision. Foremost, he cited the favorable conclusions of the Stone and Webster report, as well as positive information obtained from the NHA and engineers retained by the RFC to review the Lustron data. Goodloe defended the decision by asserting that "the Lustron house is not only a good house but is among the best, if not the best, of the various prefabricated houses that have been developed."The only aspect that gave the board pause, it seemed, was the risk that the public would not accept the house. Although cautious of the potential consequences both financial and political of backing a failure, the board, according to Goodloe, "felt like it was a risk that we were justified in taking in view of the acute need for prompt low-cost housing for veterans."[15]

Terms of the agreement included repayment over 78 months at 4 percent interest, with the first installment due on January 1, 1948. Significant elements of the loan reflected concerns among the directors regarding Lustron's ability to initiate production in a timely manner, as well as fears of "excess profit" at government expense. Consequently, the RFC structured the agreement to retain a measure of control over the venture. The guarantee of Chicago Vitreous, although limited by a complex financial formula, also allayed fears. The expertise and technical guidance provided by Chicago Vitreous reassured the RFC that an established firm stood behind Lustron. Collateral requirements included all of the machinery, equipment, and inventory of the pilot plant, as well as the assignment of all patent rights, pending patent applications, and license agreements to the RFC. Further, the agreement required preapproval of all salaries paid to Lustron directors, officers, and employees exceeding $10,000 per year.[16]

Should Lustron succeed, the RFC included a provision to maximize its position and accelerate repayment. It reserved the right, after two years, to convert the remaining balance on the loan into 100 shares of common stock for each $1,000 owed on the principal balance. Should the venture fail, the RFC secured its position as principal creditor. Additional restrictions limited Lustron's right to alter its capital structure, engage in any activity other than the production and sale of prefabricated houses and component parts, or invest more than $50,000 per year in capital improvements without the written permission of the RFC board.[17] At last it appeared that the most significant barrier to the company's future was removed. Lustron could now proceed with negotiations to secure the Curtiss-Wright plant in Columbus, Ohio, and finalize arrangements with suppliers who had been unwilling to commit to production until the loan came through.

Company Building

Strandlund celebrated the milestone in the Mayflower hotel dining room with VFW Commander-in-Chief Starr and Charles Prins during the evening of June 30. Starr and other leaders of national veterans' organizations had been invaluable allies. Strandlund reiterated the company's intentions to favor veterans for dealerships and also promised to give priority to veterans in filling an estimated 7,000 jobs at the Columbus plant.[18] Yet as the champagne flowed at the Mayflower, the owners of Chicago Vitreous, William and Emanuel Hogenson, began to develop second thoughts about their participation in Lustron. A lack of consistent communication on Strandlund's part, a consequence of his immersion in the politics of the startup process and physical separation from Chicago, prompted the brothers to reevaluate their relationship with the enterprise. An extensive review of the loan agreement and financial impact analysis by the Hogensons' attorneys and auditors alerted them to a number of potential personal and corporate hazards. The brothers were in their late sixties and planned to retire in the near future. Consultants advised the Hogensons against accepting the guarantee provision of the Lustron loan since it limited short-term liquidity, created a number of immediate tax problems, and might force a major reconfiguration of their estate plans with unknown effects on the tax liability of their heirs.[19]

Restrictions placed on the amount of profit that Chicago Vitreous could realize from the sale of frit to Lustron also raised concerns. According to the agreement, Chicago Vitreous's profits were limited to a percentage at or below the average profit margin on sales to other customers. On the surface this seemed reasonable. But the company estimated that an additional $500,000 to $750,000 investment would be necessary to rationalize the increased plant capacity required to supply Lustron with a previously agreed upon 50 percent of its frit requirements. If Lustron failed, the Hogensons would be "on the hook" for a large amount of money and Chicago Vitreous would be saddled with excess plant capacity. Further, the loan agreement prevented Chicago Vitreous from selling the Lustron common stock it would receive for development expenses without prior permission from the RFC.[20]

An additional motive behind their decision to withdraw support was "the fear of political taint," as Emanuel Hogenson later put it. If Lustron failed, questions might be raised about the high level of governmental participation in the loan. The Hogensons feared that potential "bad press" and the appearance of "guilt by association" might permanently damage the sterling reputation of Chicago Vitreous.[21] The brothers had worked too hard to build their company to see it placed in a vulnerable position at a time when they anticipated a comfortable retirement. Although the Hogensons remained convinced that the Lustron house was "a wonder" and believed in the ultimate success of the venture, they moved to terminate their association with the company.[22]

The Hogensons' attorneys informed Strandlund on August 5 of the decision to withdraw from Lustron.[23] Just as all elements of the venture were coming together, Strandlund once again had to ask the RFC to change the terms of the loan agreement. How many times could he go to the well before it dried up? Angered and frustrated by the actions of his longtime business associates and personal friends,

Strandlund nevertheless discovered opportunity amidst the debris of the broken agreement. He began to craft a reorganization plan for a "new" company that would afford him greater opportunity for personal control over the enterprise.

The period from July to November 1947 is a crucial and misunderstood era in Lustron's history. Until this point Lustron remained a subsidiary of Chicago Vitreous. The name had been changed from the old Porcelain Products Company, but not the legal relationship to the parent firm. With the Hogensons out, Strandlund resigned his position at Chicago Vitreous and created an entirely new corporate entity. Although the name remained the same, a "new Lustron" emerged during the summer of 1947.

Strandlund and his attorneys first faced the task of negotiating terms for the withdrawal of Chicago Vitreous. An agreement with the Hogensons provided for a payment of $340,000 for all patents, machinery, and inventory relating to the production of porcelain-enameled steel houses. However, Chicago Vitreous retained patent rights and license agreements for technology pertaining to the production of enameled-steel commercial buildings. The company also agreed to supply Lustron with 25–35 percent of its frit requirements with no restrictions on profit margins.[24]

The new Lustron also required a new capital structure. Incorporated in Illinois, the capital structure of the company included authorization of 1.8 million shares of common stock at $10 par value organized into the following categories: 84,000 shares sold to investors at $10 per share for a total of $840,000; 17,000 shares held as options to Hornblower and Weeks for their services; and 30,000 shares held by the company to offer as options to employees, officers, and directors. Strandlund and his wife Clara received 86,000 shares for a nominal consideration of $1,000. The RFC approved the purchase in part to reward Strandlund for his efforts and also to provide an incentive for success. The RFC held the remaining shares in reserve to satisfy the conversion privilege of the June 30 agreement. If Lustron succeeded, it could convert Lustron's debt into approximately 1.5 million shares and potentially reap a large profit.[25]

The actual dollar value of subscribed capital totaled $841,000, less $340,000 to buy out Chicago Vitreous, leaving a net invested capital of $501,000. Would the issue of equity participation—$501,000 against $15.5 million—again inflame controversy? Now only the personal assurance and business acumen of Carl Strandlund stood behind the loan. Yet the RFC board faced a quandary. Would it dare oppose the wishes of the administration as expressed in John Steelman's letter of support?

The composition of the new Lustron's board of directors offered some solace, since each member possessed proven commercial talents certain to aid the firm. Representing legal and financial interests were attorney Joseph E. Nolan of the Washington,

D.C., firm of Bell, Boyd and Nolan; Howard E. Buhse, a partner in Hornblower and Weeks; and Raymond Haynes, a partner in the Chicago investment house Wellington and Company. Representing Lustron's suppliers were Raymond Hurley, vice chairman of the Thor Corporation, the manufacturer of Lustron's dishwasher/washing machine unit; Louis Leverone, president of the Automatic Canteen Company, a supplier of kitchen and bathroom components; George Delp, president of the New Holland Machine Company, a manufacturer of Lustron's enameling machinery; and Paul O. Buckley, an officer of the Federal Machine and Welder Company, a supplier of production equipment for steel processing. Inside directors included Strandlund, executive vice president Russell G. Davis, and vice president and treasurer Fred M. Lowum.[26]

Despite significant changes in the financial and organizational structure of the firm, the White House remained enthusiastic in its support for Lustron. A number of Truman's closest economic advisors viewed the RFC as a convenient tool with which to seed the economy and forestall the likelihood of a postwar depression. But certain advisors also favored using the RFC to reward friends and punish enemies through the determination of loan recipients. Whether or not Truman himself approved of these tactics, the administration's manipulation of the RFC left it vulnerable to charges of favoritism and influence peddling. Steelman's letter expressing Truman's support for Lustron, for example, could easily be construed by opponents as political pressure. RFC support for the new Lustron indeed provided a good example of the extent to which the administration favored a liberal lending policy. Just as the Hogensons feared, the political consequences of such a policy would eventually taint the RFC, the administration, and Lustron.

Nevertheless, on October 31, 1947, the RFC and Lustron signed a new loan agreement and a lease for the Curtiss-Wright plant in Columbus. The terms were identical to the June 30 agreement in every respect except for the guarantee of Chicago Vitreous, the authorization of a new capital structure, and a provision requiring Lustron to pay its rent on the Columbus plant directly to the RFC as collateral rather than to the War Assets Administration. The plant agreement provided for a five-year initial lease, with an additional five-year option, at $426,800 per year.[27]

Strandlund also formally announced the management structure of the company on October 31. The majority of Lustron's top executives possessed extensive engineering experience. Executive vice president Russell Davis joined the company from Foote Brothers Gear and Machine Company where he had served as first vice president. Director of engineering and development Ralph H. Wise formerly served as supervisor of engineering for the Fisher Body division of General Motors. Chief engineer Robert Runyan and director of the enameling department E. E. Howe, key players in the development of Chicago Vitreous's commercial construction business, joined

Lustron with the blessing of the Hogensons. Sales vice president Raymond J. Hurley, a former vice president of the Electric Household Utilities Corporation, along with sales manager Carl Rolen, brought a combined fifty years of sales experience in the housing industry to Lustron. Vice president and treasurer Lowum and D. W. Boylan, comptroller and secretary, both veterans of the housing industry, completed Lustron's top management.[28]

Annual salaries ranged from $50,000 for Strandlund, a 50 percent reduction from his compensation as president of Chicago Vitreous, to $12,500 for the junior officers. Nearly all took pay cuts to join Lustron, a tribute to Strandlund and to their belief in the ultimate success of the product. Of the twenty-three top executives, three had worked at Chicago Vitreous, fourteen had held line management positions at manufacturing firms, two had worked in the publishing industry, and four had joined Lustron from federal agencies.[29]

With its management structure in place, the company now turned its attention to labor issues. Represented by the American Federation of Labor (AFL), the building trades had traditionally been suspicious of prefabrication and perceived industrialized housing as a threat to local autonomy and control of the construction process. Lustron had kept the AFL well informed of its intentions and established a good rapport with its national leadership, especially Richard T. Gray, president of the AFL Building and Construction Trades Department. Consequently, the AFL was willing to consider a new type of labor agreement tailored to the needs of prefabrication at the factory as well as the home site. On November 11 Lustron and the AFL announced a two-year labor agreement with three federated unions—the United Association of Journeymen and Apprentices of the Plumbing and Pipe Fitting Industry of the United States and Canada, the International Brotherhood of Electrical Workers, and the United Brotherhood of Carpenters and Joiners of America.[30]

Terms of the agreement included a guarantee of uninterrupted production of the houses at the factory and their efficient assembly at sites in the United States and Canada. Reducing the number of crafts involved in production from twelve to three lessened the likelihood of jurisdictional disputes. A pledge by union officials to advise locals of the terms of the agreement and enforce contracts if necessary also reduced the potential for on-site labor problems. The agreement was attractive to the AFL because it promised year-round employment, a benefit of prefabrication over the seasonal pattern of employment in the conventional building industry. With the guarantee that Lustron would be a union shop both at the factory and at the assembly site, the AFL demonstrated its willingness to support "the experiment of industrialized housing." The Lustron-AFL pact quickly became a model for agreements between the union and other prefabricated housing manufacturers anxious to foster a "spirit of cooperation with the building trades." Strandlund wrote that the agreement "with great segments of American labor [would be] cited far and wide as an example of fair dealing, honest concern in the welfare of our fellow citizens and 20th century industrial relations."[31]

To foster this cooperation, Strandlund recruited Frank J. Peterson as director of industrial relations. Peterson had a long history of labor relations experience, serving as an international representative for the United Association of Journeymen and Apprentices of the Plumbing and Pipe Fitting Industry before taking various posts with the federal government during the war. He was a member of the important Facilities Review Committee of the War Production Board and its successor agency, the Civilian Production Administration. "Pete" also served as a member of the Conciliation Service of the Department of Labor, which had provided him a broad view of labor–management issues across the economy.[32] Lustron's ability to recruit employees with substantial experience in government and industry was a benefit of its hybrid genesis and high public profile. In general, the company was able to attract highly talented people who shared Strandlund's view that Lustron was "special" and embraced the once in a lifetime opportunity to shape a new era in American housing.

Preparations for production at the Columbus plant went into high gear during November. As the first of millions of dollars worth of machinery and equipment began to arrive at the plant, the company announced it intention to hire approximately 7,500 workers in 1948 and set a production target of 100 houses per day by midyear. Civic leaders could barely contain their elation at the prospect of becoming "the Detroit of the housing industry."[33]

In Washington, however, the Senate Republican leadership introduced a resolution in December 1947 calling for a thorough investigation of the RFC. Expressing philosophical concerns regarding the role of the RFC in the postwar economy and suspicious of possible influence by the administration in its loan policies, Senate Republicans saw an excellent opportunity for political exploitation. A special subcommittee of the Committee on Banking and Currency chaired by Senator C. Douglass Buck (R-DE) held hearings during December and January to explore the details and motivations behind RFC lending practices. The existence of the RFC itself became a fundamental issue since enabling legislation listed June 30, 1948, as its termination date.[34]

The Lustron loan received extensive scrutiny. The committee subpoenaed Strandlund and Goodloe to elicit their perspectives on the history of the venture. Unlike his previous congressional appearances, Strandlund faced difficult questioning, especially regarding his purchase of 86,000 shares of company stock for $1,000, and battled a perception that the RFC had given Lustron a better deal without the support of Chicago Vitreous than with its participation. Committee members wondered why there appeared to be no limit to the RFC's support of this "equity-starved" venture. "That is the $64 question," Senator Homer Capehart (R-IN) intoned, "that is the crux of the whole thing."[35] Strandlund understood the intent to suggest scandal and favoritism but provided the committee little insight into the political machinations surrounding his company. "Now, political pressures I do not know very much about," he said, playing the role of struggling entrepreneur to perfection. He proceeded to present a credible defense of his firm and cleverly used the forum to emphasize the tremendous potential of prefabricated housing. "I wish you much success," Senator Buck wryly offered, "and hope it makes a lot of money, and you can pay the Government back."[36]

A more intense examination awaited RFC chairman Goodloe. Senator Capehart bluntly asked, "where did pressure come from to make you grant this loan?" Goodloe instantly objected to the use of the word "pressure," but conceded that "from a credit standpoint, [it] was not what you could call a sound loan." Goodloe did not deny that the Steelman letter, with its indication of presidential support, had guided the RFC toward approval but urged the committee to withhold judgment until the results of Lustron's first year could be evaluated.[37]

Further, Goodloe implored the committee to perpetuate the RFC and to resist using the RFC–Lustron situation as an excuse to terminate the agency. Perhaps new restrictions and greater oversight were needed, he conceded. But where else could small businesses, especially those operated by and for former GIs, go to obtain long-term credit at reasonable rates? Goodloe then noted that the RFC had loaned over $16 million to 3,924 veterans during fiscal 1947 and 1,852 of those GIs had been unable to obtain business loans from commercial banks "due to the prevailing policy of banks not to make GI loans."[38]

Returning to the controversy over the Lustron loan, Goodloe urged patience. Lustron was a complex undertaking tying to "do its best" as it navigated uncharted territory. Why couldn't the government exercise at least an equal amount of patience that any private investor would grant to a fledgling enterprise, especially one that held such obvious benefits for the nation? "I am not trying to avoid any responsibility in making that loan," he told the committee, "I can't say the Lustron project [will be] unsuccessful. On the other hand, I can't say it [will be] successful. It hasn't gone far enough. I don't know whether the loan is going to pay out, but the Government was justified in making it."[39]

Roll Up, Roll Up . . .

Although Strandlund's and Goodloe's testimony failed to allay the suspicions of the committee regarding the Lustron loan and RFC lending practices in general, it succeeded in buying time for the company. During the early months of 1948, management focused on promotion and production. Both presented formidable challenges.

Overcoming the negative connotations of prefabrication, particularly the association between prefabrication and "temporary" construction, demanded an aggressive advertising and public relations campaign.

A major objective of the campaign was to deemphasize negative beliefs that many homebuyers held regarding prefabricated construction, an understandable perspective given veterans' recent experiences with Quonset huts and given the overall quality and impermanence of wartime prefabricated housing. In contrast, Lustron cast its house as the equal of custom-built homes, but with added advantages only factory production could achieve, such as the use of high-quality materials manufactured to precision tolerances in every component of the house. The structural and maintenance benefits of porcelain-enameled steel also received consistent emphasis, and the company borrowed freely and effectively from promotional materials previously developed by Chicago Vitreous. The appeal to efficiency emerged as a prominent theme as well. Lustron's liberal use of built-ins, the cost-conscious radiant heating plant, and the automatic dishwasher/washing machine provided key points of differentiation that the marketing department could justifiably tout as unique, desirable, and otherwise available only in "upscale" houses.

Lustron's old friend Senator Ralph Flanders helped launch the company's promotional campaign in early 1948 by providing invaluable national exposure. Under the auspices of the Joint Committee on Housing, Flanders released a 185-page study entitled *High Cost of Housing*. The pamphlet contrasted the advantages of Lustron's "total package" with the frequent cost overruns and contractor "add-ons" that drove up the price of conventionally built houses. Additionally, Flanders emphasized the elegance of design reflected in all aspects of the Lustron house and celebrated the unity of form and function that had been achieved by "some of the foremost architects of the nation." The senator also pointed out that only a large-scale producer could afford to hire preeminent architectural talent to design homes for the common man.[40] Lustron found another friend on the Joint Committee, vice chairman Joseph McCarthy, although Strandlund would later come to regret it. McCarthy had adroitly maneuvered the committee's hearings to cast further public housing initiatives as "socialist." Interestingly, though, he frequently argued that government should assist private enterprise in the advance of housing production technologies: "There are those who maintain that because private enterprise has not solved the entire problem we should scrap private enterprise and socialize housing. But it seems only logical that instead of attempting to scrap private enterprise we should furnish the necessary aids to make it work."[41]

Others agreed. Lustron's national visibility attracted consistent attention from newspaper editors anxious to mine housing stories for national syndication. *Army Times* prominently featured the Lustron house in its veterans' edition, which had an international circulation. The sales department noted that "letter inquiries had doubled" in the week following the article. A number of real estate editors from papers in the East stopped by the plant on their way home from a builder's convention in Chicago in March, and the company cheerfully provided press packets, management interviews, and nourishment in hopes of facilitating good relations. The hospitality paid off, as the King Features and United Press syndicates distributed encouraging stories with pictures and descriptions of the house to over seven hundred papers.[42]

The initial phase of Lustron's promotional strategy borrowed heavily from the tactics of the film industry. Beginning in April 1948 and continuing throughout the year, the company held a series of "unveilings" of model homes across the nation. Amid searchlights, local celebrities, prominent politicians, and veterans' groups, the Lustron house debuted to an enraptured public. Favorable responses from the introduction of the original Cicero prototype had provided the company with a public relations blueprint. Now it was time to increase both the intensity and the frequency of the efforts to win public acceptance.[43]

On April 2, three trucks left the Columbus plant with a model home destined for display at 52nd Street and Sixth Avenue in the heart of New York City. The extent of the physical difficulties and expense of transporting the house highlighted a significant problem. Three trucks containing 250 crates cost the company over $4,500 to ship to New York City—about as much as the cost of manufacturing the house itself. Additionally, unloading the house in an efficient manner at the erection site demanded sophisticated logistics beyond anything conceived by company engineers at the time. These were serious issues, and Strandlund ordered his staff to focus on an engineered process for loading, transportation, and unloading.

As the trucks made their way to New York, an advance press conference alerted the local and national media, which gave the unveiling prominent features in hundreds of newspapers and on some thirty radio programs. Strandlund invited the American Society of Illustrators to cosponsor the event and arranged to donate proceeds from the one dollar per person admission charge to the United Nations Appeal for Children. The company also planned to host a benefit ball at the Waldorf Astoria to celebrate the unveiling. The door prize would be a Lustron house. Strandlund told the press that "Lustron homes mean better living for our own people," adding that he was "happy that proceeds from the home showing would go towards aiding the innocent victims of war in other friendly nations."[44]

For the first month of the exhibition in New York City, lines filled with enthusiastic homebuyers and extended "around the block," with total attendance exceeding 130,000 persons. Lustron's newsletter described it as "super-colossal" and offered *Variety*-like headlines such as "Lustron Home smash hit along Main Stem," and "Lights . . . camera . . . Roll 'em" to communicate the excitement and glamour of the event

The Lustron living room. Photo credit: Ohio Historical Society

to plant employees. Indeed the company had scored a major public relations coup. Strandlund personally escorted many celebrities and prominent politicians through the house and reported special interest among European visitors who wanted to know when the company would begin export sales. Several families waiting in line for hours in the midst of a driving rain told the press that they had made the Lustron tour the centerpiece of their vacation to the city but granted that the Empire State Building was impressive as well.[45]

As in Hinsdale, landscape design played an important role in displaying the house to best effect and reinforced the idea that this technologically advanced house fit easily into established conceptions of suburban neighborhoods. Privet hedges defined the perimeter of the front view, with mountain laurels accenting various points along the foundation. Cherry trees and lilac bushes added interest along the front and sides of the house. A bluestone terrace extended from the rear of the house, inviting visitors to explore a backyard framed with willow trees and incorporating a meandering bed of English ivy.[46]

Designer Mary Davies Gillies of *McCall's* magazine coordinated the interior presentation of the house. Several reporters commented that the generous built-in storage facilitated a "pleasing modern style" of interior design that accentuated the open

The Lustron kitchen, equipped with an "Automagic" combination clothes- and dishwashing machine. Photo credit: Ohio Historical Society

feel of the floor plan. "The home has more storage space than many twice its size," observed the *New York Herald Tribune*. Noting the ease of maintenance and economy of the entire design, the *Tribune* asserted that "[w]omen who find themselves keeping house in a Lustron Home shouldn't have too hard a time of it." New York radio personality Norman Brokenshire offered his listeners the kind of publicity that even Carl Strandlund hadn't anticipated when on-air he addressed a child, Wesley Pearce, and gravely intoned: "Young man, you are standing in the greatest single development in housing since they first put one stone on top of another."[47]

The culmination of the New York event took place in the grand ballroom of the Waldorf-Astoria on the last day of April. Over 3,000 people attended the event, and over 100,000 tickets had been sold by the Society of Illustrators for the Lustron house door prize. Noted artists such as James Montgomery Flagg, Steven Dohanos, and Norman Rockwell sold tickets, and other Society artists had rendered the model home and donated their creations to the company in appreciation of its efforts. Even the winner of the door prize provided Lustron with invaluable publicity. Mrs. Michael Toosusian of 306 West 28th Street in New York City had been down on her luck. Shortly after her recent marriage, her husband suffered a heart attack. As he lay an invalid in the bedroom of their fourth-floor walk-up apartment, Mrs. Toosusian tried to maintain their small dry cleaning and tailor shop as well as tend to his recuperation. Upon learning of her prize, she beamed "this is a dream come true" and announced plans to secure a lot near the Long Island shore where she hoped her husband would make an accelerated recovery.[48]

With the New York exhibition an unqualified success, Lustron dispatched another model home to Washington, D.C., for an unveiling on May 28. The cosponsor of the event was the company's best ally to date—the Veterans of Foreign Wars. Commander-in-Chief Starr personally assisted top management in recruiting powerful members of government to attend private tours of the house. Other VFW members served as guides for the public, distributed literature, and collected donations for the organization's rehabilitation fund. Strandlund also conducted private tours for members of major industrial and financial lobbying groups such as the National Association of Manufacturers. Even though the event was marred by bad weather, the public and the media swarmed the house. Five radio shows featured the opening, including Arthur Godfrey's popular program, and all the Washington papers ran extensive coverage in their daily and Sunday editions.[49]

Enthusiastic reception by the media and the public energized the company as the unveiling process continued throughout the summer and fall of 1948. Lustron erected model homes in Chicago (the Cicero prototype had been cannibalized for parts), Milwaukee, Indianapolis, St. Louis, Detroit, Boston, Des Moines, and, finally, in its hometown. Total combined attendance exceeded 1.5 million persons.[50] Each exhibition reinforced the concern for "context" in the display of the Lustron house and demonstrated the company's sensitivity to perceptions that a steel house could not be aesthetically integrated into traditional neighborhoods. Confident of large attendance regardless of location, the company exercised great care in choosing sites to facilitate extensive landscaping and placement of the house in "mature" settings.[51]

Lustron also participated in a number of important trade shows, including the American Federation of Labor's Union and Industry Exhibition in Milwaukee attended by over 300,000 people. Manned by Frank Peterson, the Lustron booth attracted enthusiastic visitors who clamored for brochures and asked "thousands of questions" about the company and its product. Peterson hosted a number of labor luminaries, including William Green, president of the AFL; Joseph Keenan, head of the AFL Education and Political Department; Martin J. Durkin, president of the United

The master bedroom closet space and built-in dressing table with mirror.
Photo credit: Ohio Historical Society

Association of Journeymen and Apprentices of the Plumbing and Pipe Fitting Industry; and Dan Tracy, president of the International Brotherhood of Electrical Workers. Durkin and Tracy also visited the Milwaukee model home with members of their staff. Lustron's newsletter noted with particular pride that "all classifications of the American public have been reached . . . business, labor and other segments of national life have shared in the openings."[52]

Concurrent with the unveiling of its model homes, Lustron launched an advertising campaign that focused primarily on such national magazines as *Time, Life, McCall's, Saturday Evening Post,* and *Architectural Forum* but also included direct mail and limited use of radio in large markets. Originally, the campaign featured a three-tiered approach, with general advertisements and a response card in the national magazines, followed by the mailing of a detailed pamphlet to respondents, and the creation of a referral service that recipients of the pamphlets could use to locate specific dealers in their geographical area. However, the initial volume of responses—180,000 in the first two weeks of the campaign—quickly overwhelmed the promotional department, forcing the company to drop the response card in future advertisements and restrict its focus to "home tour" layouts and "progress report" bulletins in the national print media.[53]

One advertisement alone in the April 19 issue of *Life* elicited over 50,000 responses within a week. Entitled "The House America Has Been Waiting For," the two-page layout touted the technical benefits of the Lustron product as a gateway to "a new standard for living." The advertisement clearly reverberated with the house-starved American public, literally burying Lustron's offices with mailbags filled with response cards. Harold Denton, who had joined Lustron as head of the market development department, noted that people not only wrote but "telephoned from all sections of the country. Telegrams and registered mail increased the swell . . . people came in person to the plant from as far south as Florida and from the far side of the Rockies in response to this ad."[54]

The firm scrambled to find additional staff to categorize the inquiries and hastily assembled form letters to send to prospective buyers. The responders included thousands of veterans, many of whom were enrolled in college on the GI Bill and lived in a variety of substandard housing conditions. One GI student wrote that "this home of yours sounds like the answer to my every prayer . . . do you have any more information on this wonderful new development?" Many responses also reflected the following, very American desire: "It would please me to be the first person in this city (New Bedford, Massachusetts) to own one of your houses." Writers often included photographs of their families, and some sent blueprints with ideas for variations on the house. One wanted to place a Lustron on the thirtieth floor of a skyscraper as a penthouse, and one religious soul wanted to convert a house into a porcelain-enameled chapel. Eventually the total volume of responses from the *Life* advertisement exceeded 200,000. Strandlund observed that Lustron was "on the main line," meaning that the company had tapped into a powerful social current.[55]

Lustron's publicity juggernaut continued unabated throughout the summer of 1948. *McCall's* ran a color spread on the house in their July issue. *Consumer Reports* evaluated the prototype against a wood prefab manufactured by Adirondack Homes and touted as a "dream house" by *Look* magazine. *Consumer Reports* found the Lustron "a better buy" and noted that the materials and construction were "excellent." Additionally, the United Press radio wire service ran a prepared script on the company and an interview with Strandlund that was picked up by 150 radio stations. Lustron's public relations department reported increased interest from the foreign press, including requests for information packets from journalists in Paris, Brussels, and Amsterdam. *Popular Science* ran a major story on the technical aspects of the house in June, and *Modern Industry* planned an extensive feature on the plant in a fall issue.[56]

The themes evident in Lustron's initial advertising campaign contained rich social and historical implications regarding home ownership. Equating the "American dream"

The mechanical room.
Photo credit: Ohio Historical Society

with ownership of a detached, single-family home, advertisements heralded the social benefits of the Lustron house as America's "first truly low-cost mass-produced home." Lustron informed the public that the application of new housing technologies "far removed from the technique of hammer and nails" would improve society by making home ownership "a realistic and attainable dream for the 'average Joe.'" References to the success of mass production in the automobile industry also provided an easily understood contextual comparison.[57]

The themes of personal and social accomplishment and the role of technology were nicely expressed in the opening paragraph of Lustron's original promotional brochure. The last sentence introduced another prominent promotional theme, that modern prefabrication could provide the "best of both worlds"—an aesthetically desirable house as well as a functional, easily maintained one:

> Nearly every American has the deep-down desire to own his own home . . . to put down roots and establish himself and his family as part of the life of the community. For all too many, this desire seemed destined to remain a beautiful dream—until the Lustron House made it an easily accomplished reality. Modern engineering "know-how" has made it possible. Now, a "home of your own"—at a low first cost and even lower upkeep cost—is within the reach of the average family. And, more than that, the Lustron Home is the kind of home most Americans want.[58]

Lustron conceded that overcoming public suspicion of prefabrication could not be accomplished through advertising alone. But the company believed that if the public could be "softened up" enough to elicit interest in the product, it could win them over with the excellence of the house itself. In an "interview" with Strandlund included in several advertisements, he reminded consumers that the Lustron house should never be confused "in any way" with previous types of prefabricated designs. Strandlund also tried to allay perceptions of impermanence by asserting that "while we will help relieve the housing shortage, this is not an 'emergency' or 'stopgap' project, but is planned on the long-range basis of complete customer pride and satisfaction and as a new contribution to the art of living."[59]

As the campaign progressed, innovative features of the house received greater attention. Design, maintenance, and mechanical systems were detailed in a series of "home tour" advertisements that ran in national magazines each week for several months. Special emphasis fell on descriptions of features not found in conventionally built houses of the target market segment, such as built-in storage and display cabinets, pocket doors, the pass-through space between the kitchen and dining room, the combination clotheswasher/dishwasher, and the radiant heating system. Customers were also frequently reminded that the house achieved its technological mission within the context of a "conservative, ranch-style architecture."[60]

Lustron's advertisements consistently focused on the permanence and ease of maintenance inherent in a house built from porcelain-enameled steel. Able to withstand "the coldest blasts of the North as well as the salt air of Florida," it was a house that "your grandchildren will enjoy owning as much as you will." The Lustron house also offered "a safer and more healthful environment for families" because its steel structure made it impervious to vermin and insect infestations. Ease of cleaning, fire resistance, and the fact that the house never needed repainting, refinishing, or reroofing, provided other key selling points.[61]

One innovative advertisement compiled frequently asked questions from letters sent to the company by "people who have already heard about the Lustron Home and are enthusiastic about it." The answers addressed potential "negatives" as well as positive elements about the house and specifically reminded customers that they would

Steel skeleton of the Lustron house.
Photo credit: Ohio Historical Society

have to give up some flexibility in return for permanence and ease of maintenance. This preemptive approach is best illustrated by answers to the following questions:

> Q: What about lightning?
> A: The house itself is a self-contained lightning rod.
> Q: Can I have any other floor plan or room arrangement?
> A: Not *yet*.
> Q: How do I hang pictures?
> A: By using self-adhesive hooking pads supplied by Lustron.
> Q: Can I have a basement?
> A: No, basements are unnecessary and outmoded for modern life—pioneers needed them for food storage, you have a convenient ground-floor utility room.
> Q: Can the house be taken down and moved?
> A: No, it is not demountable or portable. If you move to another location, you will want to sell and buy another Lustron Home.[62]

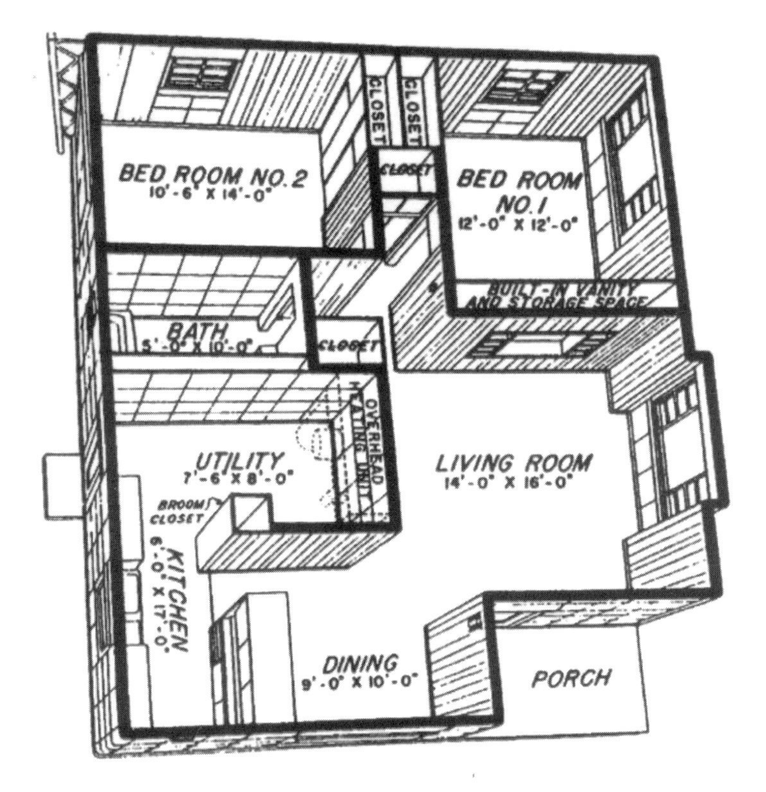

Floor plan of the two-bedroom model. Author's collection

Lustron's advertisements reflected a constant struggle to advance the merits and minimize the drawbacks of a "closed system" of manufacture. They stressed that each design element, even mundane components such as gutters and window frames, was fabricated for exclusive use on a Lustron house. Other products could not be substituted, nor could homeowners easily alter the house to reflect personal preferences or tastes. The closed system, they emphasized, also ruled out aftermarket modifications such as room additions, floor plan variation, or redecoration by painting or papering.

The company initially viewed the closed system as a marketing opportunity as well as a challenge and looked to the automotive industry for strategic cues. Alfred Sloan, General Motors' legendary chairman, had demonstrated the effectiveness of "style depreciation," a marketing strategy that encouraged customers to purchase a new car for aesthetic or status reasons long before their old car was mechanically spent. Thus General Motors introduced annual style changes and provided a ladder of consumption for status-conscious consumers to follow that began with the Chevrolet and ended with the Cadillac. Lustron embraced a similar marketing concept and planned to offer its own ladder of more sophisticated and expensive houses that would keep consumers "in the family" as they grew more affluent with age. The company anticipated that given the choice, consumers would replace houses for the same reasons as they traded cars.[63]

The closed system created a fundamental tension between self-expression and home style, on one hand, and the need for a housing design easily adaptable to mass production, on the other. The question remained, would the public accept manufacturing compromises that restricted self-expression? Lustron was convinced that it was a matter of degree. Considerable compromises were necessary during the formative years. Then, as design and production technology became more sophisticated, a greater degree of flexibility could be offered to the consumer without forfeiting the benefits of prefabrication.[64]

Manufacturing the Future

However, the formulation of a long-term marketing strategy was necessarily subordinated to the demands of preparing for full-scale production. Concurrent with the establishment and implementation of a promotional campaign, Lustron moved toward its goal of producing 100 houses per day by October or November 1948. During the first months of the year, over $12 million of equipment arrived at the Columbus plant and company engineers, with the aid of the Lindberg Engineering Company of Chicago, scrambled to set up a complex manufacturing process. The scale and scope of the production process far exceeded any previous experiment in the prefabricated housing industry. At full capacity, the Columbus plant would use more electricity than the entire city of Columbus.[65]

Lustron approached the manufacturing process with the goal of creating a "countrywide production line," beginning with the arrival of raw materials at the factory and ending with the assembly of the house on a lot. Innovative assembly line techniques and a complex system of material-handling equipment ensured a steady flow through the plant. Approximately eight miles of automated conveyers enabled the synchronization of production and reduced the need for warehouse space. In fact, the plant design relied heavily on continuous-process technology and tight coordination of processes from raw material deliveries to the distribution of finished houses.[66]

Responding to Strandlund's directive to develop a cost-effective method for transportation and site assembly logistics, company engineers led by Dick Reedy soon devised an elegant solution. With minimal outside assistance, Reedy's team pursued the goal of arranging a complete set of components on a single flatbed trailer. Working in an empty corner of the plant, they placed the wall sections of the house along

A Lustron house on one of Dick Reedy's specially designed tractor trailers.
Photo credit: Ohio Historical Society

the outer perimeter of the trailer and secured them with roof truss assemblies. This created a container into which the other parts could be loaded in reverse sequence of their need at the assembly site. Boxes mounted on the floor of the trailer held fasteners, nuts and bolts, and all the tools necessary for assembly.[67]

Strandlund was astounded at the speed and ingenuity of Reedy's accomplishment. Compared with using wooden crates and multiple trailers, this method promised to save approximately $90 million a year based on expected levels of production. The trailer package provided a number of ancillary benefits as well. Each trailer was painted bright blue and yellow and emblazoned with Lustron logos, while the components themselves projected a high-tech look that had a dramatic visual impact. Nothing else on the road looked remotely similar, and drivers reported that the rigs turned heads wherever they went. These "rolling billboards" could also reduce raw material shipping costs since trucks returning from assembly sites that were reasonably near Lustron suppliers could transport materials back to Columbus. Another potential advantage was that a large fleet of trailers could function as a virtual warehouse for excess inventory without incurring additional expenses or compromising delivery time. Reedy went one step further—he proposed piggybacking the trailers

Lustron's young executives in front of the factory.
Photo credit: Ohio Historical Society

onto rail cars to serve West Coast markets or on river barges to serve the Southeast. While he explored that possibility, Lustron negotiated with a Chicago-based leasing company to provide four hundred tractors and eight hundred trailers built to Reedy's specifications by the Fruehauf Trailer Company.[68]

This was Strandlund's company—innovative, responsive, and creative. Lustron was unparalleled in the history of the housing industry, and the sheer scale and scope of the Columbus plant demanded resourceful employees implementing a shared vision. This was a war plant in many respects—its president was a civilian war hero and most of Lustron's management team had either served in the armed forces or held important jobs in the defense production establishment. They were used to accomplishing the impossible in an incredibly short period of time. All of them believed they could do it one more time—after all, how many times did you get to be on the ground floor of the "General Motors of the housing industry"? One such believer, General Eugene Reybold, former head of the U.S. Army Corps of Engineers, joined Lustron as vice president of operations in early 1948. Reybold had directed projects valued in the billions of dollars over his career, and as Strandlund noted, "his willingness to accept a position with Lustron is another evidence of the high regard felt throughout the nation for our venture here."[69]

Aerial view of the Lustron factory in Columbus, Ohio.
Photo credit: Ohio Historical Society

Lustron established a plant newsletter and welcomed all new employees by establishing the scope and spirit of the enterprise. "We have assembled the nucleus of a staff which, I am sure, is destined to make productive history," Strandlund wrote in the first issue; "the work being done here at Lustron and the cooperative manner in which it is accomplished is enough to warm the hearts of all of us." He noted several pressing employee issues, including the lack of public transportation to the plant and, ironically, the scarcity of available housing in the Columbus area. Evidently the complexity of the cafeteria "production" line caused a number of difficulties, but this was soon organized to everyone's satisfaction. Strandlund took great care to remind employees that they were "making one of the vital and most important contributions to the American way of life. . . . It is a challenge worthy of the best in all of us."[70]

Executive Vice President Russell Davis picked up this theme, noting that Lustron "is in the throes of one of the most promising, satisfying, and important formations in the history of American industrialization . . . we are engaged in building the best homes yet offered to the American public at any price." Lustron management also consistently emphasized the service to returning veterans, noting the "disruption, irritation, instability, and even divorces" that had resulted from inadequate housing. Casting the company's operations as part of a continuing war effort served as a rallying cry for the fledgling enterprise and established a tone of urgency and patriotism throughout the organization. It was up to Lustron's employees to provide the industrial ingenuity that would revolutionize the residential construction industry "as positively and aggressively as . . . other great sections of our economy."[71]

As the plant came to life, the enormousness of the production enterprise before them impressed everyone who witnessed it. The manufacture of a Lustron house began with steel fabrication. Steel arrived in rolls and sheets via truck or by railroad car at a specially designed four-track spur. Unloaded by overhead cranes directly into the plant, conveyers fed the steel into various shearing, punching, and stamping machines. For high-volume parts such as the two-foot-square wall panels used extensively in each house, coiled steel was fed continuously into six-hundred-ton punch presses that fabricated the complete panel in a four-step process. Similar processes formed the roof panels, bathtubs, sinks, and lavatories. The punch presses "overformed," or bent, the panels to create attachment edges, a significant improvement over welding edges onto a flat panel. The overformed panels thus maintained maximum tensile strength through subsequent fabrication, enameling, and erection processes.[72]

For lower production volume parts, the "miscellaneous press department" maximized efficiency by performing several different operations on the same press. Most of the presses had combination dies that required only minor adjustments to produce a variety of parts. This innovation reduced tooling costs and saved time since operators did not have to reconfigure the presses for each production run.[73]

Structural assemblies such as roof trusses and various bracing systems were fabricated on roll-forming equipment and conveyed to welding stations where workers completed the 16,000 welds necessary for each house. Three separate "loop conveyers" held the largest structural components together and passed them by welders working simultaneously on three levels. The tri-level arrangement eliminated the need for workers to reposition their welding guns from one level to another, a time-consuming and accident-prone maneuver.[74] From the stamping, rolling, and welding operations, components traveled by overhead conveyers to the enameling department.

A comparison of Lustron's raw material requirements with the entire U.S. enameling industry provides a startling insight into the scope of operations. Based on a projected production of 100 houses per day, Lustron would consume approximately 25 percent of the frit and approximately 75 percent of the coloring oxides produced in the nation. Each day, the plant would enamel over 100,000 individual pieces in over 200 shapes and colors—by far the largest enameling operation in the world. Ceramics industry publications reflected both elation and trepidation at Lustron's potential impact.[75]

Porcelain-enameled steel panels emerge from drying ovens.
Photo credit: Ohio Historical Society

The enameling department consisted of a mill room where the ingredients for porcelain enamel were prepared and eleven enameling furnaces that operated as self-contained "miniplants." The mill room occupied 13,000 square feet and contained 42 mills that produced the enameling solution, or "slip," by grinding clay, color oxides, opacifier, and frit in a fourteen-step process. Prepared slip was kept fresh by agitating tanks that gravity-fed the solution directly to the enameling line. Each enameling furnace included a preparation, or "pickling," unit that cleaned the steel; a "bonderizing" unit that applied a protective ground coat; a dryer; an automatic sprayer that applied the slip; and, finally, a continuous process, 180-foot furnace that fired the solution at 1,500 degrees Fahrenheit. The enameling line ran at speeds from seven to twenty feet per minute, resulting in an hourly capacity exceeding 12,000 square feet of porcelain-enameled steel.[76]

After fabrication and enameling, components proceeded to the assembly department. Finished panels first received a plastic edge gasket that sealed joints between panels, then proceeded to stations where machines automatically affixed glass wool insulation to the inside of the panel. The final destination was the shipping department, where workers sorted and arranged completed parts for shipment on the custom-designed trailers.[77]

The spectacle of production overwhelmed many contemporary observers, who, with a combination of pride and amazement, expressed admiration of Lustron's "great American factory." Even the typically staid *Fortune* magazine described a "special kind of beauty" in the production process:

> The twenty-three acres of floor space are alive in one long rhythmic flow . . . the monstrous two-story-high presses stamping steel into exact shapes; the tall shiny wire baskets, floating gently, swaying like mobiles by Alexander Calder, carrying things around a couple of miles of production line; dozens of huge mixers rolling endlessly overhead . . . long steel beams coming together, laced in a pattern and then sent on a wide merry-go-round through an electronic welder that works like a thinking machine . . . the endless lines of huge trailer trucks, slowly sliding forward along a track toward the daylight.[78]

By the summer of 1948, the total value of Lustron's plant and equipment had already exceeded the combined investment in the next forty largest prefabricated housing firms.[79] Clearly, the company shouldered the burden of the entire industry in the minds of investors and the general public. If it couldn't succeed, it seemed, nobody could. As Senator Flanders observed, "if Lustron doesn't work, let us forever quit talking about the mass-produced house."[80]

But was size a curse or a blessing? Lustron's detractors already possessed compelling evidence of the dangers inherent in high-volume, narrow-market production. The "Lustron bathtub story" appeared frequently in critiques of the company's progress, and RFC chairman Goodloe later reported that his files contained "nearly 100 copies" of the same article about "the bathtub" sent to his office by those critical of the Lustron loan.[81]

Problems began when the company decided to fabricate its own bathtubs, a result of frustrating bulk price negotiations with major suppliers. The engineering department then approved the purchase of a very expensive 1,800-ton press to stamp out bathtubs from rolled steel at an estimated 35 percent savings per unit. The problem was that Lustron houses needed a five-foot, 1.5-inch bathtub, while the industry standard was five feet. The 1.5-inch design variation proved extremely costly. Amortization of the investment in the press required a production rate of at least 120,000 bathtubs per year. Even at full production, the company could only use 36,000 per year, and the size variation meant that no market existed for the excess production. Retooling of the press and redesign of the house to accept a five-foot bathtub eventually cost the company far more than the original plan to use outside suppliers.[82]

Additional problems surfaced in July 1948 as Lustron prepared to initiate "warm-up" production. Production and promotional expenses had nearly exhausted the orig-

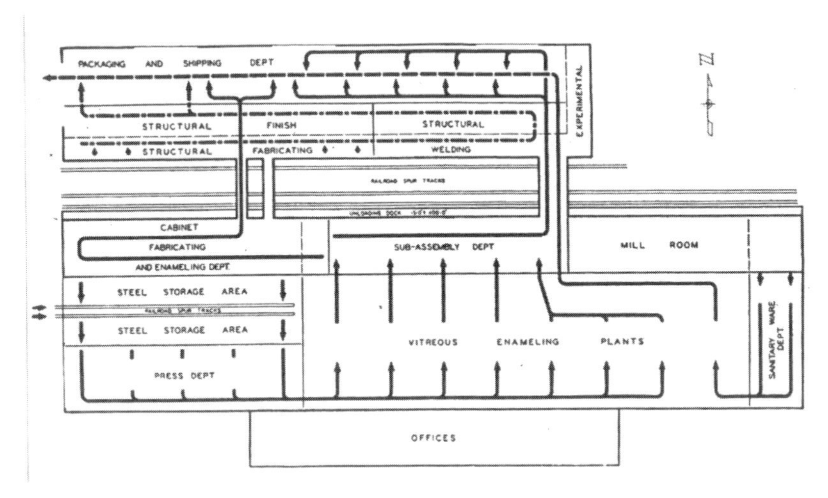

Schematic of the Lustron factory. Author's collection

inal $15.5 million RFC loan. Strandlund had promised and fully intended to raise further capital through private sources, using FHA guarantees as collateral. But when Congress revoked the FHA's authority to insure private loans in June 1948, Lustron was forced to approach the RFC for $10 million in additional capital.[83]

The RFC again retained the Stone and Webster Engineering Corporation to review Lustron's progress, analyze its management and organization, determine whether further investment was warranted, and provide estimates of costs and profits at various production levels.[84] Stone and Webster's report, issued on July 15, concluded that an additional $10 million loan would be "amply sufficient to bring the operation to volume production." Reviewing progress at the Columbus plant, the report raised no "serious apprehensions" and indicated that full production of 100 houses per day appeared to be attainable by the end of the year. The report also praised Lustron's personnel, noting the high level of experience and competence among senior management, as well as the "considerable number of promising young men."[85]

The core of the report analyzed production and sales costs and offered an overview of Lustron's competitive position. Based on a monthly production of 3,000 houses, Stone and Webster projected the unit cost of a Lustron house at $4,359. This figure represented $3,300 for materials, $382 for labor, and $677 for factory and administrative expenses. With a projected selling price of $4,800 F.O.B. Columbus, the company would realize a profit margin of $441 per house. Thus the break-even point to cover production expenses plus interest charges on the new loan total of $25.5 million would be approximately 1,000 units per month. At 3,000 units, Lustron's gross profit would exceed $1.3 million per month. "We are confident that the finished product can be sold . . . at a price that will be competitive with houses of comparable size and quality," the report concluded.[86]

Stone and Webster refused to predict the extent of public acceptance, yet it noted that over 250,000 people had responded to Lustron's initial advertisement in *Life* and that more than 4,000 people had applied for dealer franchises. Although it warned that the plant might not be large enough to meet demand, the report concluded favorably on Lustron's progress and prospects and advised the RFC to make an additional $10 million loan.[87]

Given its previous proclamations of support and the favorable report from Stone and Webster, the RFC had little choice but to grant another loan. On July 11 the RFC board approved an additional $10 million loan at 4 percent interest, although the agency required full repayment by October 31, 1949.[88] The RFC's decision was aided by congressional amendment of the National Housing Act in August 1948, which allocated $50 million to the RFC to "make loans to and purchase obligations of any business enterprise for the purpose of providing financial assistance for the production of prefabricated houses."[89] The fact that Lustron could not provide additional collateral did not deter the RFC, which agreed that the original loan conditions would suffice. The initial loan had been granted under a "defense" mentality driven by the veterans' housing shortage. The supplemental loan agreement made no mention of crisis conditions. It was simply a business loan. Increasingly, the line blurred between private control and federal involvement in Lustron.

Another critical problem facing the company was the limited availability of steel. At full production Lustron would require approximately 500,000 tons of steel annually, about .5 percent of total U.S. production in 1948. Again, Strandlund exploited his political connections. Senator Flanders requested that the Department of Commerce, through its Office of Industry Cooperation (OIC), allocate 60,000 tons of steel to the prefabricated housing industry. Lustron was to receive 40,000 tons—the lion's share of the initial allocation.[90]

The nation's steel fabricators immediately resisted. Protests to the OIC claimed that "White House pressure" had created a "smell of favoritism" for Lustron throughout the government. Dr. C. J. Rodman, president of Alliance Ware, a plumbing fixtures company, and head of the Metal Plumbing Division of the Porcelain Enamel Institute, observed that 60,000 tons of steel could be used in as many as 40,000 conventionally built houses but could produce only 6,000 prefabricated houses. "The Government's entire relations with Lustron have an extremely bad odor," cried Rodman. Why did the government continue to shelter Lustron at the expense of proven manufacturers?[91]

With $25.5 million and considerable political capital invested in the enterprise, Rodman's question seemed naïve. Strandlund and others in Lustron's top management naturally saw the relationship with the government in different terms. Executive Vice President Russell Davis praised the government for its "farsighted outlook" and called its relationship with the company "very healthful." Strandlund himself admitted that Lustron was "in the habit of getting what it wants."[92]

Various appeals from Lustron executives to the OIC sought "special treatment" to help the company obtain sufficient steel to meet the target production rate of 100 houses per day by the end of the year. The appeals were based in part on powers granted to the president under the Stabilization of the Economy and Commodity Prices Act, which Truman had signed on December 30, 1947.[93] The law gave the president authority to establish advisory committees representing major industries in an effort to encourage cooperation between government and business. The program sought to further economic stabilization by developing voluntary anti-inflationary wage and price agreements and plans for the allocation and control of scarce commodities that impacted the cost of living or industrial production.[94]

The Steel Industry Advisory Committee, which included the presidents of the nation's five largest steel companies, initially showed little inclination to allocate significant quantities of steel for the production of prefabricated houses. The "steel boys" had enough problems meeting the needs of established customers, and many remained wary of governmental pressure, especially regarding Lustron. Roy Ingersoll, president of the Ingersoll Steel Division of the Borg-Warner Corporation, agreed with Dr. Rodman that administration support of Lustron was "unwarranted" and confessed that he was at a loss to understand "why so many people in high places were continually promoting . . . such a wild project."[95]

Supporters of Lustron such as John Steelman and Housing Expediter Tighe Woods saw another possible motive within the steel industry's reticence. Woods wrote to Steelman that he suspected U. S. Steel, which had a history of interest in prefabricated housing, "will not encourage . . . allocations to Lustron; then, after it goes through the wringer à la Tucker, will buy up the company at a nice fat loss to the government." Woods affirmed his belief that Lustron was "still the best answer that anyone has been able to come up with for low-cost housing" and encouraged Steelman to assist the company to obtain the necessary raw materials.[96]

Carl Strandlund also urged Steelman to take "any action which you may see fit" and enclosed a "bulletin of progress" detailing Lustron's promotional activities. Strandlund noted that "the overwhelming response from hundreds of thousands of American homeseekers justifies the foresight shown by congressional, administrative, veteran and labor leaders in our corporation's economic contribution."[97] Steelman agreed and urged the Department of Commerce to relate to the steel committee the administration's desire to see Lustron up and running. Indeed, despite outcries of favoritism, Secretary of Commerce Charles Sawyer signed the steel allocation order in mid-August, and Lustron "got what it wanted" once again. On August 31, 1948, the Columbus plant produced its first run of porcelain-enameled steel. An avid cigar smoker, Strandlund promptly ordered some of it made into souvenir ashtrays decorated with his signature.[98]

CHAPTER FOUR

Adventures in Sales and Marketing

Houses are built to live in and not to look on: therefore let use be preferred before uniformity, except where both may be had.
—Francis Bacon, *Essays, Civil and Moral,* No. 45, "Of Building"

The Lustron factory came alive in September 1948. From the balcony of his office at the Columbus plant, Carl Strandlund could overlook nearly twenty-three acres of production machinery. He loved to escort visitors to the balcony and chuckle as they stood speechless before this spectacle of American engineering and manufacturing prowess. Even a staid, somewhat condescending reporter from *Architectural Forum* dispatched to profile Lustron admired "the throb and pound of the big presses, the clash of steel against steel, and the sigh of the blowers" that defined the rhythm of the factory. Strandlund's office contained numerous mementos reflecting Lustron's brief and tumultuous history. Behind his desk stood autographed photos of President Truman, Senator Flanders, and Housing Expediter Wyatt. Selected newspapers articles and Lustron advertisements adorned the walls, and souvenir ashtrays made from the first run of porcelain-enameled steel sat on every table.[1]

Because the company had evaded so many potential disasters, Strandlund believed that his "tempered" organization could withstand and even capitalize on adversity. He claimed 20,000 "firm" orders on the books and announced a 1949 production goal of 45,000 houses. The company designated its original offering the "Westchester," a model similar in design to Blass and Beckman's prototype "Esquire" but with numerous improvements. The model name evoked the upscale New York county, a choice intended to associate the house with established and desirable suburban communities. With pilot production underway, sales, marketing, and distribution operations evolved and expanded as the company implemented strategies to accommodate the demands of mass marketing.

Unusual among manufacturers of factory-made housing in this era, Lustron established methods for integrating its house within the context of traditional neighborhood and community settings. The careful staging of the exhibition houses had reflected this concern. The company formalized and codified its vision in the "Lustron Planning Guide," which offered specific strategies to encourage "the proper placement" of the house within "the proper environment." This reflected a holistic approach to the sales and marketing of prefabricated houses that the industry had often neglected. The adaptability of Lustron houses to suburban development represented a crucial marketing element. The company either had to take on the role of developer or forge relationships with merchant builders who were beginning to construct large-scale suburban developments on the periphery of cities across the nation.

Lustron's sales and marketing functions necessarily developed quickly. It established, refined, and extended its franchised dealer network, addressed the difficult issue of financing the purchase of prefabricated houses, and forged an innovative relationship with a mortgage-lending firm that made it easier for dealers and customers to obtain financing for the house before it was assembled on a lot. The company also created a Fleet Sales Department to facilitate the development of large-volume merchant business and began to court the military housing market. Traditionally the Achilles heel of the prefabricated housing industry, the sales and marketing functions were extraordinarily complex and nuanced due to the particularistic character of the American housing market. Lustron's top management nevertheless believed that the company "had the house Americans want[ed]" and confidently asserted that the organization stood "right on the eve of success."

Market Realities

Lustron had demonstrated that it could design and manufacture a desirable factory-built house. What had yet to be determined—by the company and by the industry—was whether such a house could be marketed successfully. Unless the company had an efficient sales and distribution organization, economies realized through mass production were irrelevant. Lustron's scale and scope revealed a confidence in its abilities to capture sufficient market share to implement its "countrywide production line" concept. Indeed, to contemporary observers Lustron personified the daring, vision, and expertise required to bring the product and the market together.[2]

Yet the unification of product and market presented formidable challenges. Management accepted new responsibilities that producers of conventional housing typically did not face. Foremost, sales and marketing systems had to be large enough to generate a continual flow of orders to rationalize mass production. Ideally, sales had to precede production to allow for efficient scheduling of fabrication and distribution.

The greater control manufacturers exerted over the sales function, the greater their control over production and distribution costs.[3] However, the market environment faced by prefabricators made it extremely difficult to rationalize sales efforts as readily as the manufacturing process.

Compounding these problems were the enormousness of the enterprise and the innovative aspects of its product. A benefit in terms of publicity and general excitement, fitting the square peg of a porcelain-enameled steel house into the round hole of long-established traditions in the residential construction industry presented unique challenges. Considering that the company was new, the product was new, the pressure from Washington unrelenting, and the level of public scrutiny intense, Lustron did an outstanding job in a remarkably short period of time.

The practical aspects of marketing a prefabricated house differed markedly from conventional homebuilding. In conventional construction, builders procured materials from a variety of suppliers and relied on subcontractors to provide the labor and expertise for each construction specialty. Depending on the type of sale, financing could be handled one of two ways. If the house was built for an existing customer, the customer paid as the work progressed with his own money, or on terms arranged with a mortgage provider. Builders thus had little of their own capital tied up in construction costs. Additionally, materials suppliers typically extended thirty- or sixty-day credit terms that carried builders financially until the customer made the next payment. If builders initiated construction before securing a customer or if they intended to market a subdivision on a merchant basis for sale upon completion, they could approach banks or savings and loans for a construction loan. Either way, the conventional builder had access to reliable sources for financing a proven product.[4]

Lustron's factory-built house required the conceptualization and implementation of an entirely different marketing and financing approach. Indeed, significant variances from conventional techniques had to be developed for the company to succeed. Because Lustron houses arrived at the site complete, dealers needed only to arrange for site preparation, foundation construction, utility hook-ups, and assembly of the house. However, this meant that the house had to be paid for all at once since the company initially operated on an F.O.B. basis. Dealers had to prearrange financing or provide their own capital. Complicating the financing process was the radical nature of the house and the newness of the company. Lenders had to be convinced of the quality of the house and the ability of the dealer to complete assembly, a process that transformed the house from chattel property on the trailer to mortgage-qualified real property.[5]

Obtaining financing for builders and customers also complicated the company's marketing efforts. Lustron had to convince lenders that its house was mortgage-qualified property, a decision that involved the consideration of numerous factors including market appeal and acceptance, location, lot characteristics, and an assessment of its value compared with conventionally built houses. The company also needed to secure approval by the Federal Housing Authority (FHA) to reach the broadest possible market. An FHA-insured mortgage reduced the required down payment since it covered a higher percentage of the appraised value of the house. This meant that the house was within reach of families lower on the income scale, a key factor in rationalizing the company's production capacity. FHA approval had a larger meaning as well by representing acceptance in the eyes of the public that had come to recognize the agency's stamp of approval as an arbiter of quality and stability.[6]

Lustron and most other prefabricated housing firms relied on a franchised dealer system of sales and marketing. Independent dealers became the middlemen of the prefabricated housing industry primarily because of the great variation in building codes, mortgage practices, consumer tastes, and real estate laws and customs throughout the nation. The franchised dealer system initially held a number of advantages for the company and the dealer. Franchising provided broad market coverage at a reasonable cost to the company since dealers assumed the financial responsibilities of distribution. Lustron could share promotional costs and exercise pricing control through the dealer contract. The sales department also believed that the franchising system furthered public acceptance of the Lustron house by associating it with "exclusive sponsorship."[7]

For Lustron's dealers, the franchise system guaranteed the right to market one of the nation's most well known commodities free of competition within a geographic territory. The system also afforded dealers the flexibility to choose which type of operation best suited their financial and operational goals. Dealers could sell to individuals before installation of the house on a particular site via a contract for delivery; they could become project developers and construct subdivisions on a speculative basis; or they could broker the house and subcontract the installation process to another firm. Lustron preferred project developers who obtained large tracts of land for subdivision and thus generated a large sales volume. The company believed this offered the best option for the homebuyer as well since the Lustron house would be situated in an area where the dealer and the company had worked together to develop a "modern community."[8]

Since manufacturers could not possibly master an infinite number of market variations in an efficient manner, dealer selection and facilitation became crucial functions. The success or failure of a prefabricator was largely determined by the ability of its dealers to generate sufficient sales volume to keep both the factory and the dealer in business. Further, the dealer also assumed responsibility for meeting consumer demands such as competitive pricing and access to financing.[9] Dealers faced important labor issues as well, including training installation crews and providing them with steady work. Workers naturally wanted assurances of steady employment before they invested their time in specialized training. Servicing houses also presented problems since dealers needed to maintain a parts inventory as well as train their crews in the proper techniques for repairing any damage to the structure or to the porcelain-enameled surfaces occurring during assembly or in the future.[10] It was a complex relationship subject to many variables. For the prefabricated housing industry, it was also largely uncharted territory.

These difficulties and other unknown factors inherent in marketing the Lustron house greatly influenced the organization and strategic direction of the company's sales and dealer relations efforts. A fundamental concept embraced by top management was that Lustron would "start big." Of course, this approach was essential to recover the tremendous investment in plant capacity and to realize economies of scale, which had been the reason for Lustron's existence in the first place. However, creating an organization capable of marketing 30,000 to 50,000 houses per year proved no easy task.

The sales organization initially included a sales vice president, sales manager, assistant sales manager, seven sales representatives, and clerical personnel. The Sales Promotion Department, which handled advertising, and a Franchise Committee, which evaluated and selected dealers, completed the organizational structure. The sales vice president's primary responsibilities included recruiting dealers, helping them to initiate operations, and assembling a field sales organization to support dealers and customers. The vice president also participated in the establishment of criteria for dealer selection and retention as a member of the Franchise Committee. Additional members of the Franchise Committee included the sales manager, assistant sales manager, the president of the company, legal counsel, and the vice presidents for accounting and finance. The Committee met weekly in lengthy sessions to evaluate numerous applications for dealerships and conducted personal interviews with finalists before granting approval.[11]

As Lustron's "point men," the sales representatives served essential entrepreneurial functions. The company expected them to develop territories by locating prospective dealers and acting as their advocate before the Franchise Committee. They also functioned as corporate liaison in all matters of dealer–home office relations and were held partially accountable for dealer performance. Sales representatives also had to devise methods of operation that suited a wide variety of circumstances depending on geographic variation or the preferences of individual dealers. It was a demanding, high-pressure job with few rules other than "succeed." The entire company depended heavily on the efforts of the sales representatives to recruit and facilitate good working relationships with dealers.[12]

Lustron began to assemble its dealer network shortly after securing the first RFC loan in October 1947. In consultation with marketing experts, the company focused on the eastern United States, particularly the Northeast and the Great Lakes states, sometimes known as the "prefab belt," where market research indicated the highest level of consumer acceptance of prefabrication. Lustron projected that it would need approximately 100 dealers immediately and decided to use geographically exclusive marketing territories. By mid-1948, Lustron had received over 6,200 applications for dealerships "with every state and a number of foreign countries represented."[13]

The company also adopted basic policies for dealer selection. Foremost was the preference for established, well-financed businessmen with construction experience since a significant initial capital investment was necessary to rationalize an exclusive territory. Indeed, sales were contingent upon a large "up-front" investment by dealers because the company was not initially in a position to finance dealer start-up costs. "We want people of standing and recognized ability in the communities throughout the country . . . [w]e want people who both understand and will practice our philosophy," wrote sales vice president Richard Jones. Yet dealers had to possess the initiative to accommodate regional and local variations in site selection, financing, and customer acceptance. Lustron expected dealers to develop their territories to full potential and initially left them free to implement tactical decisions without consulting the home office. The dealer was essentially a customer of Lustron operating under certain conditions, but in some respects the relationship resembled a partnership.[14]

The Dealer Franchise Agreement provided the legal basis for dealer relations and defined in detail the limits of exclusive territories. The only exception to exclusivity was in the case of "fleet sales," a special marketing category in which the home office reserved the right to handle all sales to high-volume customers such as large-scale suburban developers or the military. The Agreement also set payment terms and capital requirements, outlined the factory warranty against defects for one year after delivery to the site, and reiterated the company's commitment to providing support for crew training and marketing assistance.[15]

Many initial dealer contacts occurred during the national "unveilings," Lustron's term for the model home publicity tour of 1948. Some prospects even harassed Lustron receptionists in their fervor to campaign for a franchise. A formal approach to dealer recruitment began in March and April 1948 when Lustron hired several sales representatives to serve in a recruitment and liaison function. Additionally, a full-page advertisement in *Life* during April encouraged interested prospects to contact the factory for a dealer information packet.[16] The sales department reported a very high level of enthusiasm among prospective dealer candidates. So confident of the opportunity, some requested more houses than the entire projected production run for the year. "There is not the slightest doubt in their minds about the public acceptance of this house," sales manager Carl Rolen wrote, "and they stand ready to put up large sums of money on this belief."[17]

Stanley Crute, who organized C-B Homes in late 1948 as Lustron's exclusive dealer in Connecticut, provides a good example of the type of individual who became interested in the company. A civil engineer, Crute operated a general contracting firm before the war and served as state director of the Office of Price Administration through 1945. He describes his attraction to Lustron:

> After the war I couldn't stand the price of lumber. I didn't think any kind of housebuilding involving the inflated price of lumber would be a sound operation. I decided to see what had happened in prefab. When I saw the Lustron prototype in Hinsdale, it looked to me like the first soundly engineered house in which it could be fairly said that machines had replaced men.

C-B Homes also hired Gregory Bardacke, a veteran of the prefabricated housing industry. He had served as a consultant to Wilson Wyatt at the National Housing Agency, and had aided Lustron as an independent consultant during the company's labor contract negotiations with the American Federation of Labor.[18]

Other dealers and their employees reflected similar technical backgrounds. Svend Ericksen, the New York City dealer, was a partner in the Gotham Construction Company, a commercial builder. Joseph Miele, the New Jersey dealer, was a civil engineer, while Chicago-area dealer Kuhne-Simmons Company included several construction engineers. Partnerships between people with construction experience and other local businesses likely to profit from an association with Lustron represented another promising approach. Prospective partners included real estate agencies, ready-mix concrete firms, and merchant builders. A number of dealers expressed concerns about "static" from entrenched residential construction interests, but interestingly saw themselves as part of Strandlund's "educational force" of Lustron constituents who would "change the misunderstandings" about the role of industrialized housing in the postwar marketplace.[19]

Although Lustron benefited from much goodwill and positive anticipation from its dealer community, the company was initially ill-prepared to manage the intricacies of this complex economic relationship. A fundamental issue inherent in the franchise marketing system was geographical exclusivity. Initially, every Lustron dealer held exclusive rights within a certain territory. All of New York City was represented by one dealer, as were the states of Connecticut, New Jersey, Florida, and large areas of Michigan and Wisconsin. Difficulties arose, however, when several prospective dealers found it difficult to raise the necessary capital to launch a dealership on such a scale. Further, some of the groups or individuals who could raise sufficient capital were initially unwilling to risk their resources on an unproven product. However, a large territory, effectively capitalized and exploited, offered greater flexibility because size enabled dealers to compensate for marketing or acceptance problems in any one locality.[20] When prospective dealers turned to Lustron for advice on these issues, the company initially had little to offer. One of the early flaws in its dealer relations program was that the company lacked data that could have informed decision making on issues such as how many total dealers would be needed, or what variables most influenced the determination of an exclusive geographic region.

Lustron expected that a period of adjustment would be necessary to formulate responses to these problems and assured dealers that it would pursue further market research. The company did, however, offer training classes for dealers and their installation crews beginning in June 1948. To prepare dealers for their promotional role, the company also offered seminars in sales and advertising techniques and provided them background information about the inception of the firm and Strandlund's vision of its future. However, the company remained convinced that the house would sell itself and that problems of installation and financing presented the greatest challenges to individual dealers.[21]

In November 1948, Lustron held its first "dealer convention" at the plant. For two days, fifty-three men from twenty-two dealerships attended a variety of sessions including technical discussions about the house and the plant, overviews of the prefabricated housing market, and seminars on advertising strategy and financing issues.[22] One of the most important aspects of the convention was the introduction of the "Lustron Profit Unit Plan." Fundamentally, the plan addressed the necessity for sales to exceed the pace of factory production. With the slogan "The Key Is Three," the company introduced dealers to the "profit unit" concept. The economics of the Lustron sales approach required that dealers erect a minimum of one profit unit, consisting of three houses, each month. Profit unit potential in any given region would thus eventually govern production and pricing policies and serve as a basis for the determination of geographic restrictions on franchises.[23]

If dealers failed to fulfill their part of the equation to order and erect the maximum number of profit units their territories could absorb each month, the entire operation from raw material purchases to final sales would be disrupted. Thus, Lustron advised dealers to erect houses before they were sold in order to initiate and maintain momentum, reduce financing delays, and maximize sales potential to anxious customers. As sales vice president Richard Jones observed, "there are many people who will buy a house that is 'ready to live in' who could not otherwise be sold or who are in such a hurry to buy that they will not wait." With over twenty years of marketing and sales experience in the housing industry, Jones was instrumental in the formulation and implementation of Lustron's profit unit plan.[24]

The profit unit concept was based on the assumption that dealers could realize consistent profits only if they maintained systematic, sustained activity. Controlling labor costs proved especially important. The plan provided a method to assure the efficient utilization of labor by keeping assembly crews continually employed. Additional advantages included better workmanship and quality control through steady employment of trained crews, better labor relations, and the protection of dealer investment in crew training. In short, systematizing the installation process promised to transfer the economics of mass production to the home site.[25]

Lustron urged dealers to "start at once" erecting houses for sale after completion. The company also introduced a dealer finance plan to encourage immediate action and outlined profit scenarios based on different types of transactions. Assuming a dealer cost of $6,800 for a two-bedroom Westchester and allowing for geographic variances, Lustron illustrated the advantage of the profit unit approach. Under the finance plan the company advanced $6,800 per house, and the dealer invested $2,200 for site preparation and labor costs. A minimum profit of $1,000 per house was assumed, although geographic and market variations could easily push that figure much higher. Any profit made on land sales was not factored into the example. Lustron estimated labor costs at $2.30 per hour based on 350 hours of assembly for each house. The company estimated a profit potential of at least $3,000 on each profit unit of three houses, assuming a four-man installation crew and a $7,000 initial dealer investment.[26]

Theoretically, dealers could start conservatively with one or two profit units for a relatively small investment then expand quickly as they gained experience. However, this would be difficult and potentially uneconomic if all sales were to individuals at different locations within a sales territory. Lustron management recognized that profitable operations depended on a proper balance between individual sales and "group" sales. A group sale, for example, containing twenty-four houses, could potentially return a $24,000 profit on a $56,000 initial investment, excluding land profits.[27]

Thus, economies of scale for dealers necessarily dictated a focus on large developments. But would people want to live in a Lustron-only development or accept the Lustron house in a "traditional" neighborhood setting? Many of the nation's largest contracting firms had initially given the house a cool reception. Indeed, one of the most attractive features of factory-made housing was that customers could have it wherever they wanted, and large-scale developers existed only in metropolitan areas where the market supported mass developments. The Lustron approach promised to realize the same economies of scale found in large suburban developments but also to make them available in smaller cities and even rural areas where conventional volume building did not typically occur.[28]

House, Lot, and Neighborhood

The company addressed the potential for Lustron-only developments and the larger issue of prefabrication and community setting in a lengthy "Planning Guide" that it distributed to dealers in late 1948. Its creation evidenced an understanding of the context in which dealers would operate as well as a realization of the need for a holistic view of housing, one which had to reach beyond the factory shipping dock if the company were to survive and prosper.[29] The introduction to the Planning Guide reminded dealers that "the product which you merchandise is essentially a *house, lot* and its *environment*—the Lustron Corporation provides the *house*—the dealer and the customer provide the *lot* and *environment.*" Significantly, the company asserted its responsibility and leadership in educating and assisting dealers in the planning process. "Good planning is good Lustron business," the Guide intoned as it set about illustrating "how Lustron's 'New Standard for Living' [could] be extended to the outdoors."[30]

The first part of the Planning Guide addressed site planning for Lustron houses on individual lots held by customers or dealers. Elements of location and physical characteristics of lots received special emphasis. "Select lots located in a neighborhood of owner-occupied homes—pride of ownership results in better maintenance of yards and buildings," the company advised. A chart comparing favorable and unfavorable aspects cautioned dealers to locate houses near schools, churches, recreational centers, amusements facilities, and shopping centers. Access to convenient transportation, fire and police protection, and public utilities also served to enhance the marketability and retained value of the Lustron house.[31]

The Guide also emphasized the need to avoid locating houses near "unsightly areas" such as dumps or industrial districts and to avoid streets bearing heavy or fast traffic. Locations where adjacent property values were lower, or where "long established adverse trends" of neighborhood decay were evident, also merited rejection. Above all, the Guide cautioned dealers to make sure that sites were "located on the 'right side' of town."[32] The physical characteristics of lots also played an important role in controlling costs and assuring a favorable setting for the house. Since the Westchester was 35-feet wide and 35.6-feet deep, the Guide recommended a minimum lot size of 60 by 120 feet. Heavily wooded or severely sloped lots were ill suited due to increased clearing and foundation costs, although the Guide advised selection of "lots having one or more existing trees—they will add to the overall value of the property." Lustron also cautioned dealers to avoid irregularly shaped lots, especially where side lot lines were at an acute angle with the street, and to avoid using deep, narrow lots.[33]

To assist dealers in identifying suitable sites, the Guide outlined an "Action Plan" with a checklist that surveyed aspects of lot selection such as utilities and improvements; distances to schools, churches, and shopping centers; taxes and assessments; and a price survey of neighboring properties. The Action Plan had the dual purpose of providing dealers with a method for quick field surveys of potential sites and later as a list to match prospective customers with lots suited to their neighborhood and aesthetic preferences. The company advised dealers to focus their attention first on "areas of subdivided lots which have been left as a residue from past periods of subdivision expansion." Typically well-located with existing water and sewer connections, these areas held many cost and time benefits since they were "ready to go." Dealers only had to locate and market them as "qualified Lustron sites."[34]

Next, the Planning Guide offered advice on the proper placement of the house on the lot and suggested numerous landscaping schemes designed to display the Lustron house to best advantage. "When the house is properly placed on the lot, rooms will be well related to suitable exterior areas and make for maximum convenience and use," the guide instructed; "plan the lot and house as a unit."[35] The separation of public and private areas became a key planning objective. The public and service areas consisted of the front yard, approach walks, and driveway. The private area was the rear yard, which included the garden, play and recreation area, and space for drying laundry. The Guide advised careful planning of the private area, because "the purchase of a Lustron Home releases the owner for leisure." Freed forever from the mundane tasks of painting and redecorating, the Lustron owner needed ample space in which to grow things, "putter around," or enjoy the children safely at play.[36]

The final section on individual sites included detailed landscaping examples. Because "the mere planting of trees and bushes is not landscaping," the Guide encouraged restraint, for "a small amount of plant material carefully chosen and organized can supply a dramatic effect." The company also suggested using curved walks and diagonal arrangements of gardens and flowerbeds to create the illusion of space. Fencing or hedges around the back yard also served to delineate the private area. Most important, the Guide reminded dealers to consider the entirety of the landscape design as a crucial element of the "Lustron lifestyle."[37]

The second part of the Planning Guide, entitled "Lustron Developments," addressed site selection and planning issues for subdivisions consisting solely of Lustron houses. Again, the Guide stressed the importance of proper location, defined as areas near schools, churches, and shopping centers on the "right side" of town:

"Remember, the Lustron Home should be set in a sound economic surrounding upon a well-ordered arrangement of land and ground forms."[38] Here was an expression of the key to successful integration of prefabricated housing and a demonstration of how conventional thinking about planning could be used to bring a truly radical house to the market.

The company urged dealers to consider broader issues of development beyond "the drafting of street and lot lines." The transformation of raw land into building lots represented the difference between a wholesale and a retail product. Consequently, dealers had to pay special attention to "image" issues such as the placement of Lustron houses in relation to the natural environment, roads, surface utilities, and other Lustron houses. "This must be done in good taste," the Guide admonished, for "through variation in setback line, orientation, color, and plant material, it is possible to develop outdoor spaces, vistas, and a variety of frontages and enclosures which can enhance the entire relation between the outdoors and the interior of the Lustron house."[39]

Effective planning of a Lustron subdivision included lots designed to conform to the topography of the land and the creation of greenspaces such as playgrounds and parks within a subdivision. Setbacks of open land around entrances to the development also provided space for shopping centers and sites for schools. The Guide also recommended long blocks and the creation of cul-de-sacs to discourage through traffic. Regarding lot size and density, it encouraged dealers to focus on determining the maximum number of "saleable lots," rather than on the greatest total number of lots. As the Guide explained, "unsold lots of odd shape and size become the infestation points of future blight and neighborhood decay." Minimum recommended lot size was 60 by 120 feet, with a net recommended density (after a 25 percent reduction for land allocated to streets and recreational areas) of 4.3 Lustron houses per acre. Houses also had to be grouped "with the minimum resemblance to arrangement in rows."[40]

As with individual sites, effective landscaping played a crucial role in creating the ideal Lustron development. The Guide offered specific instructions for creating planting schemes that would screen undesirable views, deflect noise from heavy traffic areas, and emphasize variations in house placement. Liberal use of flowering trees such as hawthorne, dogwood, flowering crab, and cherry served to attract prospective Lustron buyers. Planting schemes also should not emphasize individual houses or attempt to imitate older developments by using "quick growing but short-lived varieties."[41]

The Guide also offered dealers a core of "packaged" planting designs for individual lots that could be alternated throughout a subdivision. Each design came with a list of primary and supplementary planting materials and a chart that provided detailed information about the physical characteristics and maintenance requirements of the suggested flora. By offering dealers the "services of a landscape architect," the company encouraged the creation of a "unified harmonious planting composition" for all Lustron neighborhoods.[42]

Color selection was another planning factor unique to the Lustron development. Since the company offered only four colors—"surf" blue, "dove" gray, "maize" yellow, and "desert" tan—the Guide presented several examples of proper aesthetic arrangement. Rejecting the idea of simply alternating the four colors, it advised working from the center of a block outward and evaluating the color selection of houses across the street as well as next door. It did not rule out placing houses of like color side by side, as long as dealers remained sensitive to creating "harmonious groupings" within blocks. Recognizing the challenges of working with only four colors, the Guide informed dealers that the factory was moving quickly to develop a wider range of color choices.[43]

The final section of the Guide offered a number of "good vs. bad" examples of subdivision planning that encompassed all of the elements previously outlined. The emphasis on Lustron-only developments reflected an increased sensitivity to the need for high-volume sales in order to achieve economies of scale on the distribution side of the profit equation. Emphasis of aesthetic aspects in such detail also revealed an understanding of the importance of public acceptance of the Lustron house in settings that held the greatest profit potential for dealers. The rationalization of mass production could not be accomplished without a strategic and tactical understanding of factory, dealer, and consumer needs and desires.

Financing and Fleet Sales

As factory production moved beyond the pilot stage during December 1948 and January 1949, the sales department faced a difficult period of adjustment and growth. Now was the time for the company's best efforts if sales were ever to match production potential. One goal had been achieved with the recruitment of 100 dealers. However, sales fell short of expectations. Ironically, this was exactly the opposite of what the company anticipated. Top management had believed that the limiting factor for Lustron would be production capacity, not sales potential.[44]

The intensification of the company's sales efforts included a redefinition of strategy and a structural reorganization of sales department personnel and functions. One of the most important additions was the establishment of a Finance Department to assist in locating financial institutions willing to provide interim financing for dealers and mortgages for customers. Another essential function of the Finance Department was to secure FHA approval of the Lustron house, a necessary prerequisite for securing widespread private financing and an effort that paid off quickly when the FHA granted Lustron provisional approval in December 1948.[45]

Close coordination between the Finance Department and the sales representatives also established a functional linkage to improve dealer relations. A reevaluation of the dealer finance plan, implemented through the sales representatives, extended additional credit to dealers for installation costs and reimbursed them for expenses above factory estimates. Since installation costs varied from region to region, dealers found it difficult to advertise a set price for the house. With greater flexibility built into the financing plan, dealers felt more confident in advertising set prices for Lustron homes since their profit was easier to determine.[46]

Lustron itself, however, could not afford to finance all of its dealers. For the dealer finance plan to work in the long term, the company had to secure external financing and limit its participation to serving as a clearinghouse or agent in the search for dealer credit. The Finance Department envisioned a "Lustron Acceptance Corporation" to provide interim financing for dealers and ease their burden of carrying the cost of the house until customer financing could be secured. A successful plan would thus stimulate the flow of credit and focus it where it was most needed. Only then could the company's large production capacity be rationalized.[47]

The Finance Department negotiated an innovative arrangement with the Galbreath Mortgage Company of Columbus, Ohio, during the first months of 1949. Though complicated, the advantages of the Galbreath-Lustron plan can be seen in the following scenario. When a dealer sold a Lustron house, the customer applied for a mortgage. If successful, the customer's bank agreed to pay the dealer when the house was completed or when installation reached a certain stage. The innovative element began at this point. The customer and the dealer then executed a mortgage assignment that authorized the customer's bank to pay Galbreath a portion of the mortgage money to cover the cost of the house to the dealer. The dealer then sent the mortgage papers to Lustron along with an application for interim financing and a promissory note made out to Galbreath. When Lustron granted approval, it released the house for shipment and forwarded the mortgage documentation to Galbreath. Galbreath then immediately paid Lustron for the house. Finally, when the customer's bank was satisfied with the progress of installation, it sent a check to Galbreath covering the advance and interest on the interim loan.[48] The successful implementation of the plan held benefits for each party. Buyers of prefabricated housing could more easily obtain mortgages; dealers secured a method to carry their costs; Lustron received prompt payment; Galbreath received interest for its services; and banks could participate in a reliable method to tap the prefabricated housing mortgage market. The plan was also significant as a model for the prefabricated housing industry as a whole, since creating and maintaining an effective distribution network presented tremendous challenges for every firm.

The creation of the Financing Department represented a positive response to difficult problems. Lustron also created several new departments to further its marketing efforts. The Service Department assumed responsibility for dealer relations other than financing. Its mission and functions included the establishment of channels of communication between the factory and dealers, the quick resolution of dealer complaints, and the presentation of sales and marketing programs developed by the company. The Service Department also encouraged mentoring relationships among dealers with the hope that more experienced dealers would pass along advice.[49]

The Service Department initiated an important change in the function of the sales representatives as well. Lustron now sent an experienced representative with the first order placed by a new dealer and recruited new sales representatives to monitor and assist dealers during the assembly process. This represented a differentiation of function for the sales representatives, who initially had responsibility for both sales and the house installation process for specific regions. Dealers now had greater access to factory expertise during the start-up phase, and considerable burden was relieved from individual sales representatives.[50]

Another important addition to Lustron's sales organization was the Fleet Sales Department. This department operated independently of the dealer network and remained the only type of sales activity directly and exclusively implemented by the corporation itself. The Dealer Franchise Agreement exempted fleet sales from territorial exclusivity. Lustron did not specifically define what constituted a fleet sale in the Agreement, but evidence indicates that it meant sales to merchant developers, the armed services, or rental builders. The focus on fleet sales represented an attempt to retain corporate control in a market that held tremendous potential.[51]

Because military sales held special promise, the Fleet Sales Department focused its marketing efforts on procurement officers. Top management also realized that involvement in the military housing market could provide a powerful rationale for the continuation of government financial assistance. In December 1948 the Navy granted Lustron a contract for sixty houses to be erected at the Marine Corps base in Quantico, Virginia. Lustron outbid six conventional prefabricators on the basis of lowest cost per square foot of useable space. Realizing the significance of this opportunity, the company assembled an "elite crew" culled from factory workers and construction engineers to guarantee timely installation of the houses. A key factor in the Navy's interest in prefabrication was the high labor costs of conventional construction in a remote location. Lustron's factory-made houses demonstrated an important competitive advantage in this market that Lustron could exploit over both conventional construction and prefabricators using conventional materials and producing only parts of houses in a factory.[52]

Major General Lemuel Shepherd, Jr., Commandant of the Marine Corps Schools at Quantico, praised the company in a series of letters to other commanders interested in the Lustron house. Shepherd reported that he was "more than satisfied" with the houses, particularly "because their attractive architecture, their trim appearance, and their varied colors do so much for the appearance of the [base]." Further, the Commandant relayed positive reviews from various constituencies. Engineers liked the house because of its durability and efficient design; the quartermaster liked the house because its efficient insulation reduced fuel costs; the maintenance officers liked the house because it was not subject to rot or termites and presented no fire hazard.[53]

Officers' wives told Shepherd that they loved the house because it was easy to clean, contained large amounts of storage space, and was virtually impervious to assaults by "lively youngsters." Indeed, he mused, the most attractive feature of the house from the husband's perspective was the automatic dishwasher. The overall advantages of the Lustron house led Shepherd to conclude that

> those of us who have studied this problem at some length feel sincerely that the Lustron Corporation has opened the door to a tremendous future. I, myself, believe that the techniques used in building these houses will be perfected as time goes along, so that with reduction in housing costs there will be a corresponding rise in housing standards throughout the Country.[54]

Developing fleet sales in the private sector presented different challenges and opportunities. The company relied on field sales representatives and dealers to refer fleet prospects to Columbus and expected the Fleet Sales Department to assist potential customers with the integration of Lustron houses into their plans for suburban developments. A good example of fleet sales development was the "Park Forest Project." Lustron began negotiations with American Community Builders, Inc. in January 1949 for the sale of 2,000 Lustron houses to be placed in a new subdivision near Park Forest, Illinois, a new planned suburb of Chicago rendered famous by William H. Whyte in the 1956 bestseller *The Organization Man*. After months of wrangling over a price structure, the parties signed an agreement on August 3 that offers insight into a number of marketing issues inherent to fleet sales.[55]

Lustron knew that high-volume sales were essential to the success of the enterprise, especially during its formative years when its dealer network was not yet prepared to handle full-scale production. Further, Lustron sought relationships with developers in part to obtain experience in land planning, site preparation, landscaping, and marketing techniques necessary "to promote the use of our house in large project developments." Testing public acceptance of the house in such a setting also served corporate interests. Additionally, the relationship with American Community Builders, one of the largest developers of project housing in the nation, would be influential in attracting other developers to the Lustron house.[56]

Lustron agreed to provide American Community Builders with a high level of service during the one-year term of the contract, including special factory training for American Community's workmen, foremen, and supervisors, and a team of "on-call" factory representatives to assist with any difficulties. Lustron also guaranteed the houses against all defects for a period of one year after the date of delivery. In return, American Community agreed to abide by Lustron's contract with the American Federation of Labor and to permit Lustron fleet sales personnel to observe its methods and procedures concerning land use and planning, site development, landscaping, marketing, and customer financing.[57] The Park Forest project represented an important initial effort to establish a favorable reputation as a volume supplier of housing and to develop marketing and planning savvy that would feed symbiotic relationships with large developers.

By the summer of 1949, the Fleet Sales Department negotiated additional contracts with the military and private firms. Military sales included a development contract with the Army, which tested Lustron houses for cold weather applications in Alaska and Greenland. Contracts with the Army at Fort Dix, New Jersey, for 300 houses and with the Air Force at Westover Air Base in Massachusetts for 1,400 houses evidenced Lustron's steady penetration of this market. The Strategic Air Command also approached Lustron for bids on 2,000 houses for use at eleven air bases throughout the nation. In the civilian market, initial contacts with Parkbrook Homes, Inc. of Cleveland, Ohio, produced an interim agreement to supply over 3,000 houses for a rental housing project in the Cleveland suburbs. As Strandlund previously noted before various congressional committees, the Lustron house was ideal as a rental property due to its durability and ease of maintenance.[58]

Considering the complexities involved, the company had created and refined its sales and marketing approach with remarkable speed. The Sales Department expanded from fifteen persons in September 1948 to over seventy persons by mid-1949. Over the same time span, Lustron increased its dealer network from seven to 185. Further, the company demonstrated a growing sophistication regarding differentiation of sales functions and the necessity of a new organizational structure to assist dealers and customers. The addition of the Finance and Service Departments were important steps in defining corporate responsibility for difficult marketing issues. Employee morale in the sales department remained strong, and several internal publications mentioned that "Lustron widows"—spouses of Lustron sales personnel—

were anxious to discuss topics other than the company or its product at the dinner table.[59]

Lustron learned quickly that no product, however great the demand, sells itself. The company's genesis in the midst of a housing shortage initially produced a tendency to underestimate the intensity of effort required to market a factory-made house, and overconfidence in the ease of surmounting obstacles such as local building codes and mortgage availability. By August 1949 these attitudes had vanished. Lustron realized that its long-term survival depended on innovative solutions to complex problems in marketing and dealer relations. However, such solutions required time to formulate and implement, and it remained unclear whether Lustron's primary creditor—the federal government—would allow the company to survive long enough to develop the necessary expertise.

$11 Million and More Questions

Strandlund claimed that the reorganization of sales functions, the focus on dealer training and service, the interim financing arrangement with the Galbreath Mortgage Company, and the commitment to high-volume sales demonstrated a "total market approach" unique among prefabricators.[60] He announced in January 1949 that the company would reach its break-even point of 1,000 units per month and would achieve its target production rate of 3,000 units per month within the year.[61] Neither Strandlund's friends nor his enemies ever accused him of lacking optimism. However, as the year's events unfolded, even the "rock-like confidence" of Lustron's president began to wane. The company faced two fundamental, enduring problems. First was the generation of sufficient capital to move beyond the pilot stage and toward a profitable level of commercial production. Second was the expansion of the dealer network and fleet sales efforts to a level that rationalized large-scale production. Lustron faced a vicious cycle. A low-production volume meant a high unit cost, and a high unit cost reduced the potential market for the house, which in turn called into question the viability of large-scale manufacture. The situation became more acute in Lustron's case because political pressures threatened to cut off access to capital and deny the company time to overcome initial manufacturing, distribution, and dealer relations problems.

Citing increased production and marketing expenses, the company raised the price of its house in January. The Westchester now cost dealers nearly $7,000 in the large urban markets of the East Coast, well above the initial desire to provide the nation a "good $5,000 house." With an estimated consumer cost, excluding lot, of $8,000–10,000, depending on region, the Lustron house began to creep out of the "middle market" so necessary to sustain plant capacity. In 1949, the federal government defined as middle class a household with approximately $3,500–4,000 in annual income. However, with veterans' annual income at approximately $2,500 annually, few could expect to obtain credit for a $9,000 house.[62]

Even though the initial price exceeded what most veterans could afford, Strandlund maintained that Lustron's first house should not be a "bare bones" offering. With public acceptance at risk, he wanted to provide a well-equipped, midlevel house with lots of "extras." Once acceptance was achieved, the company could then offer an entry-level model that lower-income families could afford. Strandlund also argued that lower-income buyers would benefit as second-time Lustron owners moved up the model line, leaving their homes for the economic class below. For Strandlund, "style depreciation" became a method of providing housing to lower-income families.[63]

However, speculation about the potential secondary market for Lustron houses would have to wait as the company again confronted a financial crisis. Responding to a shortage of operating capital, in February 1949 Lustron requested an additional $7 million loan from the RFC. The RFC board approved the loan on February 21 but required full repayment within six months. Lustron's total debt to the federal government now stood at $32.5 million, and the lack of sufficient cash reserves placed the firm in danger of defaulting on the original $15.5 million loan. Operating losses for the first two months of the year exceeded $1 million.[64]

The RFC's position was becoming increasingly problematic. It had consistently defended the Lustron loans, but its precarious political situation meant that further congressional scrutiny would surely be forthcoming. The directors were beginning to lose their taste for public assaults on their fiduciary judgment and winced at the prospect of opponents using the Lustron loans as a justification for eliminating the agency altogether. Some directors began to formulate scenarios for extracting the agency from the Lustron enterprise while saving their personal reputations. Although Strandlund could not have known at the time, the agency that facilitated Lustron's birth was beginning to consider infanticide.

Faced with few other options to ease the financial crunch, Strandlund announced a "temporary" layoff on February 22 affecting 400 of Lustron's 3,200 employees, blaming "the need to retool and streamline certain parts of the manufacturing process." The affected workers, who had been fabricating roof trusses largely by hand, faced a one- to two-month furlough until new tools, dies, and production machinery could be installed to complete the automation of the roof truss department. Just three weeks later, Lustron announced the layoff of an additional 400 employees. Strandlund again cited "manufacturing bottlenecks" and delays in the installation of automated machinery as the reasons for the layoffs.[65] Production line imbalances such as these were not unusual for such a large and complex plant, but they proved particularly troublesome since they were widely publicized. Their timing also worked against the company, as the need to solicit additional capital again brought Lustron's critics to the fore just when the company needed breathing space.

Between January and May 1949, Lustron produced an average of 130 houses per month, well below the thirty-five per day break-even target. However, summer brought renewed hope, for in June the plant achieved a record of forty-two houses in one day and averaged over 100 houses per week by the first of July. Although Lustron by this time had produced only 1,298 houses, many in the company sensed a turning point.[66] "We are very close to a breakthrough," Strandlund reported in the company newsletter.[67] Yet operating losses continued to mount. Lustron reported a loss of nearly $1 million for the month of May, which forced the company to seek an additional loan from the RFC.

It was not an opportune moment to approach the increasingly embattled agency. In the summer of 1949, the RFC faced mounting political pressures, especially among conservatives in Congress who continued to question the necessity of federal capital allocation in a peacetime economy and who regarded the RFC's lending practices, especially the Lustron loans, as unwise and potentially corrupt. Conservatives raised a number of embarrassing questions that refocused national attention on Lustron during hearings on legislation affecting the RFC and the prefabricated housing industry.[68]

The most important issue for Lustron was the extension of Section 102 of the Housing Act of 1948, which authorized the RFC to make loans and purchase equity obligations of any business to assist in the production of prefabricated housing or housing components. The authorization was due to expire on September 1, 1949. Among the proposed amendments to the 1948 Act was the authorization of an additional $50 million for the production of prefabricated housing and $25 million to assist dealers with interim financing arrangements.[69]

During hearings before the Senate Banking and Currency Committee, RFC director Harvey J. Gunderson offered a cautious defense of the Lustron loans in much the same manner as his colleague John D. Goodloe had done in early 1948. Gunderson told the committee that he was "reasonably satisfied" with Lustron's prospects for reaching the break-even point within the year. Despite his knowledge that Lustron planned to request an additional $3–5 million to "tide them over," Gunderson remained convinced "that [the RFC] should support the company until [it is] certain whether it is a success or a failure. The importance of this venture is, if it works, that you will get cheap housing."[70]

Gunderson insisted that additional funds were necessary to expand Lustron's distribution network and to support interim financing arrangements for dealers. As he observed, "it is a lot easier to build the houses at the plant than it is to get them out to where the customer wants them. Unless the distribution system is built up to the point where anyone . . . anywhere in the country, can talk to a dealer and get a house, the program will not work."[71] A lucid assessment, yet several members remained skeptical that the American public would embrace porcelain-enameled steel houses.

J. William Fulbright (D-AK), chairman of the committee, expressed "grave doubts" about Lustron's potential market and questioned "whether it [was] wise to continue spending $1 million a month. That is a lot of money . . . I would hate to come around next year and find that we have $50,000,000 invested in it without any appreciable change for the better." Fulbright also focused on the issue of public acceptance. "I have only seen one of them," he said, "but it sort of reminds you of a bathtub." No, Gunderson replied, the first ones looked "a little like hotdog stands." However, he quickly added that Lustron was concentrating its efforts on developing more attractive color schemes with the help of industrial designers and architects. Responding to Fulbright's query about the overall prospects of the company, Gunderson said that he thought that it had "about a 50–50 chance" of success. Nevertheless, the RFC director recommended continued support for Lustron until the government could be "completely satisfied of the likelihood of success or failure for prefab housing ventures."[72]

Before hearings began in the House, Strandlund and Representative Frederick Smith (R-OH) rekindled an old debate in the press over the level of private equity capital invested in Lustron. The company, Smith observed, appeared to be "the RFC under another name." Strandlund, though, saw "nothing abnormal" in the relationship and argued that since the loans originated under the Veterans' Emergency Housing Act, Lustron did not have to contribute any private capital. "The transaction," Strandlund said, "should not be judged by conventional ideas." Smith curtly replied that "there's a limit to how unconventional the RFC can be in lending $34,000,000 or more of public funds."[73] Yet Strandlund held his ground, asserting that "it would be tragic" if the congressional inquiries adversely affected Lustron's opportunity to borrow additional funds. "I am not afraid of the inquiry as such," Strandlund told a reporter from the *New York Times;* "there is nothing wrong or unwholesome that the investigation could uncover."[74]

In front of the House Banking and Currency Committee, Strandlund faced a stern assault from Representative Albert Cole (R-KS), who reminded him that Lustron's total losses to date exceeded $14 million. Although Strandlund admitted that additional loans were necessary, he assured Cole that "nothing has gone wrong with the project but the timetable." Cole then announced that he had learned that the RFC was prepared to lend another $14.5 million to Lustron, which would exhaust the remaining funds allocated to manufacturers of prefabricated housing under the Housing

Act of 1948. In effect, Cole noted, the "RFC may have to put all of its prefabricated housing eggs in one basket." Strandlund replied that he had no control over RFC lending practices but maintained that once Lustron "has found its feet and proved itself" it could easily obtain sufficient private capital investment to pay off its full debt to the RFC in a very short period of time.[75]

Cole then addressed a blunt question to Richard C. Dyas, chief of the RFC Housing Branch: "I want to inquire whether the Reconstruction Finance Corporation is now satisfied with the loan. Is it a good investment?" "That," Dyas replied, "is a question of philosophy. As a loan—and that may be the wrong name for it—of course it cannot be supported by any consideration of security in the sense that an ordinary loan would be . . . I think you have to consider this whole project as a pilot undertaking. I do not believe that anyone in the RFC could say that it is satisfactory or unsatisfactory. There are no yardsticks."[76]

Cole also wondered how the RFC could have interpreted the National Defense Section of the RFC Act so broadly as to include the original Lustron loan under its provisions.[77] Dyas argued that the intent of the Lustron venture was to provide housing for returning servicemen that could be considered "a move on behalf of the national defense." Cole responded, "I hope in the future that the RFC doesn't interpret the statutes so broadly." Dyas again reminded the committee of the enormous scope and innovative character of the venture and asked for additional time before rendering harsh judgments that could imperil Lustron's future.[78]

Despite the attacks in Congress, the RFC agreed to extend four additional 60-day operating loans to Lustron during the summer of 1949. Three loans of $1 million each were executed on July 7, July 15, and August 10. The final loan, executed on August 30, totaled $2 million. Lustron's indebtedness to the federal government now stood at $37.5 million.[79] Despite negative press attention aroused by the hearings and the loans, economic pressure prompted Strandlund to announce on August 31 the layoff of 700 factory workers due to "continuing production line difficulties."[80] His assertion that the company was "right on the eve" began to be interpreted by the public as a prelude to failure rather than success.

Although congressional inquiry had not prevented the RFC from extending additional credit, the agency grew increasingly sensitive to criticism about its relationship with the company. The national media, once a firm ally of Lustron, reflected many of the concerns of congressional opponents and gave the firm considerable negative coverage. *Time* magazine's coverage of the hearings included an article entitled "Bathtub Blues" that played up Fulbright's remark and concluded that Lustron "had been far from a success."[81] Reflecting its conservative nature, *Fortune* chronicled various aspects of the company's history in an article entitled "That Lustron Affair," which implied scandal and favoritism at the RFC. "It's what happens when government starts handing out 'risk capital,'" the author asserted; "this is the first time the government has appropriated, in peacetime, any considerable sum for *private venture capital.* Does anyone in his right mind want the government to decide who deserves it?"[82]

These criticisms prompted the RFC to retain the renowned management and engineering consulting firm Booz, Allen & Hamilton to begin a detailed evaluation of Lustron's operations.[83] At the same time, Strandlund and his management team undertook a similar evaluation during the summer of 1949. Continuing difficulties with the automation of several departments created additional production delays and placed more jobs in jeopardy. Increasing complaints from dealers regarding assembly problems with the Westchester model also forced the company to reassess the house within the context of production efficiency.

Reengineering the House

The manufacturing bottlenecks and site assembly problems had the same origin. The Westchester was essentially identical to the original prototype Esquire, which had been designed with market acceptance as the first priority. Architects Blass and Beckman had placed less emphasis on whether the design could be efficiently mass-produced since neither they, nor the company, knew exactly how each component would be fabricated. Consequently, numerous manufacturing difficulties inherent in the design emerged when commercial production began. Since time and monetary pressures had prevented a wholesale redesign of the prototype, Lustron now faced the consequences of those circumstances.

Dealers complained that "you could paper the walls with the assembly blueprints." For example, installation crews had to hand assemble roof trusses and spend an inordinate amount of time fitting polyvinylchloride plastic gaskets around each two-foot-square panel. The design prevented a higher level of factory preassembly that could save dealers time and money at the assembly site. Indeed, many dealers reported that assembly of the Westchester took approximately 1,200 man-hours, more than three times company estimates of 350 man-hours. Few dealers could absorb the higher labor costs, both for training and overtime, which resulted from inefficiencies inherent in the design. Although the number of hours required to erect the house dropped significantly as assembly crews became more experienced (one dealer reported a decrease from 1,211 hours for the first house to 436 hours for the tenth house), most dealers experienced less of a reduction because they split their experienced crews up to train new crews. Yet this proved problematic because each new dealer would face the same costly "learning curve."[84]

Realizing these "designed-in" problems would continue to plague operations, Lustron sought the aid of Carl Koch, a prominent Boston architect. Koch had been a senior architect at the National Housing Agency during the war and served as a master planner for the U.S. Air Force. His firm, Carl Koch and Associates, had extensive experience in community planning and in prefabrication techniques. Assessing Lustron's dilemma, the firm initiated a fundamental reassessment of the product line with the goal of reengineering the house to achieve two basic objectives: streamlining the manufacturing process by reducing the number of component parts and emphasizing a greater degree of factory preassembly; and enabling the introduction of additional models, both at the low end and the high end of the market, without increasing production complexity. Koch also provided a valuable ideological outlook, stressing the interrelationship and interdependency of each element of the "Lustron process" and challenging the company to work toward a truly unified approach to product and market.[85]

Koch began the process by "starting at the end"—the assembly site—and "working [his] way back" to the factory floor in an attempt to understand each element of the design and its relationship to manufacturing efficiency. As Koch observed, "the design of a house is in no way separable from the ways in which it is to be made."[86] He identified several fundamental problems. First, the prototype "had never been architected in a real sense." Although Koch praised Blass and Beckman's work as "remarkably good and workable," he argued that production delays would remain inevitable due to the excessive number of parts in the house that forced workers to "hitch the house together by hand" at many points in the manufacturing process. Drawing on the often used analogy to automobile production, Koch reminded Lustron that a multiplicity of parts was not a concern for automobile companies because their dealers did not have to assemble the final product. The cost and headaches of site installation remained a crucial issue for the company since dealer profit was closely tied to the ease and pace of the assembly process.[87]

Koch's team began a thorough analysis of Lustron's operations. Primary challenges included improving the efficiency of the company's current machinery and sequence of production, reducing the number and complexity of component parts, and expanding the product line without increasing plant capacity. The introduction of a three-bedroom model also received priority. Market research predicted great success for a three-bedroom model since such a house, whether prefabricated or conventionally built, was rare in Lustron's market segment. Further, a steady increase in birthrates since the end of the war meant that the successful introduction of a three-bedroom unit would serve an important consumer desire and sustain the future growth of the enterprise.[88]

Such plans, however, were predicated on a reassessment of the entire design concept. Koch initially focused on a key aspect affecting both production and site assembly. He noted that the system of two-foot-square porcelain-enameled steel panels wasted the opportunity to use the steel as a structural element of the house. The original design employed steel framing members, much like the studding in a conventionally built house, with the panels affixed to the members. To prevent leaks, each panel also required a polyvinylchloride gasket surrounding its edge. Every exterior wall thus used a large amount of polyvinylchloride and required labor-intensive operations at the factory to guarantee proper installation. This design also added to installation costs since workers had to attach hundreds of panels. As Koch observed, "the house weighed about twelve tons altogether, half of it holding the other half up."[89]

Koch's revision called for an increase in the size of the porcelain-enameled steel panels from two-feet square to two-feet wide by eight-feet long. Company engineers initially balked at the idea because they thought that the enameling ovens could not be modified to accept the enlarged panel. However, with creative assistance from consulting engineers from the Massachusetts Institute of Technology, the engineering department accepted the revision. Koch also suggested that the panels be rolled out and cut to length in a continuous process, rather than cut and punched intermittently. Since the panels now served as load-bearing members, his team devised a method for "ribbing" the steel, or bending it back upon itself, which increased the strength of each panel. Ribbed edges also created an improved joint, which eliminated the need for polyvinylchloride gaskets while increasing weather resistance. Although the panel redesign required additional equipment, retooling, and a revised production layout, Lustron approved because Koch demonstrated that the new system would pay for itself within several months due to savings in labor and material costs. Additionally, dealers would benefit from a more efficient design that required less site labor, and customers would benefit from cost reductions and improvements in the performance of the house.[90]

Koch's team also assessed the design of the roofing system. The roof employed panels formed to simulate Spanish tile, and fit together without the use of gaskets. Concerned with potential leakage, the company assigned an engineer to test different sizes and shapes of panels in an effort to increase weather resistance. Koch thought the panels looked awful. However, Lustron resisted his plans for a major redesign because the present system proved cost effective. The company, moreover, had invested nearly $5 million in equipment to produce the panels and was unwilling to consider fundamental changes. "We resolved this problem," Koch mused, "by deciding that the roof didn't look so bad after all."[91]

Koch's team also argued for a new window design. The Westchester used four different window sizes. Koch developed a modular unit that was completely interchangeable with the new wall panels. This design enabled a multiplicity of window placements and combinations adaptable to different models, yet retained the modularity essential to reduce manufacturing costs. Impressed with Koch's effort, Strandlund suggested adding electric windows as a unique marketing feature. Koch considered the idea frivolous until he discovered through a cost analysis that Lustron's economies of scale made an electric unit only slightly more expensive than conventional window assemblies. Koch also suggested eliminating the front bay window since it could not be adapted to the modular design concept. Besides, he found the bay window "incongruous" with the overall appearance of the house.[92]

Interior spaces also received close attention. Assessing the floor plan of the Lustron house, Koch discovered that by increasing the size of the house from twenty-nine by thirty-seven feet to twenty-nine by forty-five feet, the company could offer four different interior plans and three different three-bedroom models. Each of the floor plans could also be inverted, which doubled the options available to customers and produced a marketing advantage without using additional component parts.

Most significantly for dealers, Koch's team reduced the number of separate components on the delivery trailers from over 3,000 to thirty-seven by advocating preassembly at the factory of the window sections, storage units, and plumbing walls. Koch predicted increased quality control as well as cost savings, noting that experienced factory workers created a better product than "builder's crews slithering around in rainy weather."[93] With fewer parts to assemble at the site, dealers could erect the houses quickly and in sufficient quantities to rationalize their operations. Successfully implemented, this idea solved one of the most vexing and costly problems facing dealers.

Reflecting on the efforts of his team, Koch considered the reduction of component parts as his greatest contribution:

> In general we had reduced the complexity of parts, and as a result the size and complexity of machinery that would fabricate them. We had substantially increased their interchangeability. We had much increased the variety of plans from which the individual buyer might choose, and reduced the price he should have to pay for it.[94]

Clearly, Koch had shown Lustron the path to increased manufacturing efficiency and produced sound ideas that addressed fundamental problems. But financial restrictions prevented the company from immediately implementing the more ambitious aspects of Koch's plans. Strandlund envisioned implementation of the necessary production changes by mid-1950, with the "second generation" models reaching the market in 1951. Yet the company needed additional models immediately. During the summer and fall of 1949, Lustron engineers developed several new models, each a variation of the original Westchester design.[95]

Front and rear view of a Lustron three-bedroom model with attached garage.
Author's collection

The new models included a three-bedroom Westchester and a Westchester "Standard," a version stripped of certain features including the automatic dishwasher/washing machine and some built-ins. Next was the Meadowbrook, slightly smaller than the Westchester, in both two- and three-bedroom models. The Meadowbrook also did not include the dishwasher/washing machine but retained all of the built-in storage units. Strandlund saw the Meadowbrook as his "Chevrolet" model "fashioned to suit middle income folks who want the many features of Lustron at an affordable price." One- and two-car detached garages provided another timely addition to the product line. The "garage package" employed standard two-foot-square porcelain-enameled steel panels affixed to traditional wood framing furnished by dealers. Package options included plans for breezeways, patios, carports, and screened-in porches that also used porcelain-enameled panels in combination with conventional materials "to give unlimited variety to Lustron Homes."[96] Rounding out the revised product line was a much needed entry-level model, the Newport. Designed "for buyers who can afford only a minimum investment, or who want to build several houses for rental income," the Newport fulfilled Strandlund's longstanding promise to serve the lower end of the market by providing a model "for the average Joe."[97]

Problems on the Home Front

The development of new models clearly placed Lustron in a better competitive position. However, continued production delays largely undermined its advantage since the company could not quote delivery dates to dealers and their customers. In the New York City area alone, over $200,000 in down payments remained in escrow as production lagged. Several customers who had placed down payments as early as July 1948 and remained without a house initiated legal action against the company and various dealers.[98] Although they frequently vented their frustrations to dealers, most customers chose to wait. Ironically, the company had apparently overcome public resistance to prefabricated housing—people had visited the prototypes, read the national advertisements, and desperately wanted a Lustron house. As sales representative J. C. Tedlie observed, "the enthusiastic acclaim of the public in general was gratifying . . . and exceeded that of dealers."[99]

The key factors of timing and price represented the primary barriers to Lustron's success. By mid-1949 the worst effects of the postwar housing shortage had passed for Lustron's target market. Conventional builders had also overcome materials shortages and stood ready to undercut many potential Lustron sales by dealers. Although shortages remained in the low-income market, even the entry-level Newport proved too costly for this segment. Indeed, the average price for all Lustron houses on a lot was $10,500, well beyond the reach of lower-income families.[100] Such price levels also placed limitations on dealers, who were encouraged by the company to construct houses before securing customers. Most dealers simply refused to follow this directive. Preliminary indications of a recession in the national economy for late 1949 and 1950 also convinced dealers to proceed cautiously.[101]

The inability of the company to advertise a fixed nationwide price for each model became an additional marketing disadvantage. Since land, foundation, transportation, and labor costs varied, the final cost to the consumer could differ significantly within the same city, or even within the same block. Ironically, Lustron had once claimed that the separation of house and lot pricing would be an "educational force" in the housing market by producing "price stability and predictability for customers."[102] Instead of price stability, Lustron customers faced numerous variables that affected the final cost of the house. This "unknown" became a serious disadvantage and one over which the company could exert little control.

Further, dealers did not necessarily share the high-volume, low-margin assumptions of the company. Many did not believe that cost savings achieved by prefabrication should be passed along to the customer. Considering the speculative nature of dealer operations, including efforts in site preparation and variations in labor costs, most argued for "charging what the market will bear."[103] Although Lustron attempted to impose controls over dealer pricing and profit margins, these restrictions proved unreasonable and ultimately unenforceable from the dealer perspective. How could the company possibly determine adequate profit margins for dealers throughout the country? There were simply too many variables. For example, foundations in colder regions had to be deeper, which drove prices up by as much as $700 per house. Regional variations in labor costs also meant that Lustron houses would be more expensive in large urban areas—the very markets the company needed to penetrate to rationalize plant capacity.[104]

The pricing issue illustrates a fundamental weakness in the franchise dealer system for marketing prefabricated housing. The company consistently sought greater control over its pricing structure to establish a competitive position in the marketplace. On the other hand, dealers, especially those just beginning operations, resisted all limitations on profit and rebelled against any form of centralized control over pricing elements that varied at the local level.

In contrast, the postwar success of site prefabrication, best illustrated by the efforts of William Levitt and Sons, indicated the marketing advantages of purchasing materials in volume and focusing on efficient organization of the building process at the site to achieve production economies. Levitt thus retained control over the price of the final product by avoiding high distribution costs and other price uncertainties associated with the franchised dealer system. Consequently, the cost of a Lustron house in the New York City area averaged $9,500 without a lot, while houses and lots in Levittown sold for approximately $8,000.[105]

Lustron defended its marketing strategy by insisting that a high-volume dealer operation based on the concepts advanced in the Profit Unit Plan would in the long run produce greater profits than conventional construction or on-site prefabrication. The company also reminded dealers that it provided them with national advertising and passed along savings from production and purchasing economies, two factors that reduced dealer overhead costs to levels well below those of conventional builders.[106]

By mid-1949, however, Lustron dealers had grown increasingly frustrated with the company. Many voiced concerns about production delays and reported an attitude of inflexibility and high-handedness from the Columbus plant. Dealers frequently commented that in areas where they sought guidance and advice, such as a compilation of "best practices" on a number of key topics, the company was unforthcoming. Yet Lustron was willing to exert control and influence in areas where dealers most wanted to maintain flexibility. Dealers believed they understood the vagaries of their markets far better than the company and desired the flexibility to find the proper pricing and marketing strategies that would best cultivate their territories.

Complaints ranged from philosophical differences to practical shortcomings that impacted dealers' ability to serve customers. They reported "missing parts, wrong plans, damaged parts, [and an] oversupply of insulation" in many shipments. Parts for one house had to be cannibalized to finish another, and several dealers had to lease warehouse space to store excess insulation. Dealers also complained that the tolerances indicated in the plans and in the installation manual were far too closely engineered. "We people in the field don't even have rulers as fine as 1/32 of an inch," one dealer chastised a Lustron engineer. "We production engineers know mass production, it's got to be done our way," the engineer replied. The tension between factory engineers and dealers remained a source of continuing friction, and the company seemed unwilling or unable to respond in a satisfactory manner.[107]

Lustron faced additional obstacles that inhibited its development on the local level. Restrictive building codes and conflicting requirements of FHA state offices continued to present a minefield over which prefabricated housing firms trod uneasily. Regional variations in FHA mortgage insurance approval policies exasperated the company and its dealers by creating numerous financing and installation delays. The FHA in Indiana, for example, insisted on an overhead light in the bedrooms while the Ohio office did not. In Tennessee, the FHA required a wall and a door between the kitchen and the dining room, while most other state offices had no objections to Lustron's open floor plan. The time and energy invested by the company and individual dealers to obtain variances from these requirements added to the cost of the house and lowered dealer profit margins.[108]

Although Lustron provided legal assistance and strategies to overcome local barriers, dealers faced day-to-day problems on their own. Tailored to the needs of conventional construction, existing building codes challenged all dealers. The sheer number of individual building codes—600 in New Jersey alone—presented formidable political and economic barriers. The code problem was even more acute in Lustron's case due to the radical nature of the steel framing structure and the use of porcelain-enameled steel exterior and interior panels. Existing codes made no provisions for these materials, and each variance had to be addressed on the local level.[109] Formulated to govern conventional construction, the codes had little relevance to factory-made housing. The prefabricated housing industry actively campaigned for code reform, arguing that codes should assess the performance of building systems rather than judge prefabrication within the context of traditional materials and construction technologies.[110]

Although many codes granted local officials authority to approve new systems and materials, few defined proper procedures for assessing prefabricated construction. Local officials often balked, even in the face of overwhelming evidence indicating that prefabricated designs exceeded performance standards established for conventional construction. The enforcement of anachronistic codes also became a convenient method for conventional builders and their political allies to prevent the introduction of prefabricated housing into specific regions. Even prefabricators working with conventional materials drew the attention of entrenched construction interests. Lustron's porcelain-enameled steel construction naturally faced intensive scrutiny and suspicion. One of the key selling points of the house, the combination dishwasher/washing machine, also ran headlong into codes prohibiting clothes and dishes to be washed in the same unit. In an effort to forestall resistance, the company urged dealers to submit plans and specifications of the Lustron house to local officials at the earliest possible date. The company also supported, through the Prefabricated Home Manufacturers' Institute, a national lobbying campaign for code reform.[111]

Indeed the company expended considerable resources fighting for code variances, but with mixed results and ongoing frustration. For example, many localities prohibited copper plumbing or required the use of locally manufactured pipe. The city of Chicago did not permit residential construction using steel exterior walls. Other localities required basements. In Lustron's hometown of Columbus, local codes also severely limited the use of steel for residential buildings.[112]

Lustron adopted a proactive approach where possible. In St. Louis the company used the local press to exert pressure for code reform. The city council voted to amend the codes and adopt performance-based evaluation after the *Post-Dispatch* published a series of editorials criticizing the city for maintaining highly restrictive codes that effectively banned prefabricated housing. In Detroit, Strandlund personally testified before the city council in support of a Lustron owner who had violated a deed restriction requiring brick construction. The original owner of the property testified that had he known in 1909 that porcelain-enameled houses would someday be available, he would not have restricted the property. Ultimately, after an unsuccessful appeal to the Michigan Supreme Court, the Lustron owner agreed to construct a brick facade around his house.[113]

Even in its own backyard, the company had to wage a public relations battle with residents of an upscale Columbus neighborhood, who argued that Lustron houses would lower property values . Strandlund emphasized Lustron's "quality engineered construction" and argued that his house would enhance rather than degrade the neighborhood. "The Lustron house should not be pigeon-holed," he asserted, "nor seen as a cheap substitute or a slum clearance project."[114]

Although battles over codes continued, the company believed that it would ultimately overcome the problem. Since the Lustron house was a standardized product, dealers needed to wage a successful fight for acceptance only once to open up their markets for future expansion. Yet neither the company nor its dealers could predict the amount of time needed to gain acceptance on a nationwide level. Another possibility was simply to avoid areas with established codes and focus instead on developing rural and suburban locations where Lustron could exert its influence to establish favorable codes.[115]

In addition to the marketing challenges presented by local building restrictions, issues surrounding the transportation of Lustron houses presented serious barriers to the development of a nationwide distribution system. Sole reliance on the "rolling warehouses"—now painted bright blue and yellow for increased visibility and in homage to Strandlund's country of birth—became increasingly problematic. Strandlund had favored the Columbus plant in part because two-thirds of the nation's population lay within a 700-mile radius. However, transportation costs using overland trucking increased sharply beyond a 300-mile radius. Although costs decreased if a tractor could retrieve a trailer from a nearby Lustron site and return it to Columbus with raw materials, this could not be assumed. Relying exclusively on this method of transportation thus severely limited Lustron's ability to serve a broad market.[116]

One potential remedy that the company vigorously pursued was "piggybacking" its trailers onto flatbed railroad cars. However, railroads based their rates on a unit cost basis. For example, each house shipped to the West Coast from the Columbus plant incurred a prohibitive rate of nearly $1,700. Lustron attempted to negotiate a commodity rate instead, which reduced the cost to approximately $600, on the grounds that commodity rates would apply if the company broke the house down and shipped it in carloads of like parts. But the railroads refused to budge, and the company remained without a long-term strategy for cost-effective national distribution.[117]

Financing difficulties for customers and dealers remained an important obstacle to success as well. Prospective customers faced a conservative banking industry that viewed prefabricated housing with great suspicion. Bankers had no frame of reference for evaluating Lustron's factory-made house. Further, during the immediate postwar years banks and savings and loans, hedging against inflation in the high-demand housing market, tended to appraise all houses below construction costs. Prefabricated housing, bearing the burden of a new product with unknown public acceptance, received even lower valuations. One savings and loan executive noted that his firm did not "even invite anyone to sit down and discuss a loan unless he has 40 percent of the purchase price for a down payment . . . and the property he is buying must not be too old or too ultra-modern."[118]

FHA approval of the Lustron house increased the confidence of mortgage lenders. But local FHA offices could insist on variations that in turn caused lenders to undervalue the house. For example, in determining the amount of mortgage insurance granted on a Lustron house, many FHA offices did not factor in the automatic dishwasher/washing machine, the built-in cabinetry, or the radiant heating system. This meant that Lustron customers had to provide larger down payments based on inaccurate FHA valuations.[119] Although Lustron had proactively addressed the difficulties of financing a prefabricated house with the Galbreath agreement, many dealers complained that finance charges cut too deeply into their already thin profit margins. Further, due to a lack of established valuation and resale guidelines for Lustron houses, the consequences of customer default and repossession could not be fully quantified. Most dealers refused to enter into any financing agreement without knowledge of their potential liability.[120]

Ongoing problems with the dealer relationship reflected a fundamental weakness in Lustron's sales organization. The company expected dealers to compensate for a lack of vision and organization at the home office. Lustron clearly delegated too much responsibility to dealers to determine their method of operation and needed a more aggressive approach in solving problems common to all dealers. Although the company was fearful of imposing tactical limitations on methods for dealer operations, a clear strategic vision would have created an atmosphere of mutual cooperation rather than an atmosphere of mutual distrust. Evidence indicated that dealers communicated far more with each other than with the Columbus office.[121]

In general, at a time when difficult problems demanded innovative solutions, Lustron approached dealer relations in an ad hoc manner. The company did not even create a comprehensive marketing plan to present to prospective dealers. Because so many unknowns remained, many excellent prospects, especially large-scale developers, proved unwilling to risk an association with the company. Due to a lack of coordination, the company "reinvented the wheel" each time it faced difficulties with new dealers.

Additionally, a high degree of turnover and conflict within Lustron's top management further strained dealer relations. By October 1949 seven vice presidents and three directors had resigned. The Sales Department also suffered key defections at a time when the company needed continuity of personnel to achieve stability.

Consequently, lines of communication and responsibility frequently broke down or became overloaded. At one point all thirty-eight sales representatives—Lustron's field troops responsible for dealer liaison and recruitment—reported to one sales manager. Severe cuts in Sales Department personnel also took a toll on morale. During September 1949 the company laid off half of its sales force. Only fourteen remained.[122]

Reeling from these setbacks, Lustron initiated a major reorganization in October 1949. The primary goals included developing a strategic sales policy and implementing short- and long-term solutions to the most pressing marketing problems. The company also created a new organizational superstructure, designated the Business Organization Section (BOS), to oversee the restructuring. The BOS immediately created two new departments, Building Codes and Export Sales. It also assigned a team of accountants to create a Lustron Acceptance Corporation with the goal of revolutionizing financing for prefab customers much as the General Motors Acceptance Corporation had done for consumer financing in the automobile industry during the 1920s. As Strandlund observed, "customer financing of a home should be reduced to the simplicity of installment buying."[123]

Lustron also decided to identify profitable dealers and concentrate on assisting them in any way possible. Out of a total of 220 dealers, only sixty reported steady profits.[124] The company believed it had little choice but to ignore marginal or unprofitable dealers for the time being. Further, since almost all of the sixty were located within a 400-mile radius of the plant, the company decided to limit its scope of operations and focus the efforts of the remaining sales representatives within that range. As one sales memorandum observed, "every profitable dealer should have a single, known point of contact with the factory at all times. This should not only expedite service, but should eliminate the necessity for day to day contact between top executives of Lustron and individual dealers on matters that should not require executive attention."[125]

Additional elements of the reorganization included a greater emphasis on fleet sales. Lustron instructed sales representatives to provide special assistance to dealers most active in high-volume sales and implemented a quota system with bonus payments to encourage and reward sales representatives who developed high-volume sales projects. Military sales also received priority, and the company considered establishing a subsidiary to develop and service the military housing market. Another change included a revision in the dealer franchise agreement that required dealers to sell a fixed number of houses each month to retain their status. "The volume of business Lustron must have cannot be achieved by 'custom selling and custom financing.' Houses should be built before they are sold," stated the policy circular to dealers.[126]

The company also reconsidered other elements of the dealer relationship. First, it ended the practice of granting exclusive marketing territories. "As many dealers should be appointed in any specific locality as are needed to exploit the full market potential," a BOS directive asserted; "cost and competition will fix economic boundaries of operation." Lustron also revised the criteria for dealer selection. Dealers would now be chosen from among existing suburban developers, real estate firms, or mortgage finance firms to exploit their experience and knowledge. The company set a long term goal of "at least 1,000 dealers" during the BOS reorganization. Lustron also agreed to sell to any builder or developer interested in using Lustron houses for project development whether franchised as a dealer or not, provided "his local reputation is good and he agrees to accept the responsibilities covered in the franchise agreement as to training, warranties, and erection."[127]

Lustron recognized the need to reestablish confidence by adopting a more aggressive sales and public relations policy. Though a sound idea, the company could not afford the services of top-level advertising agencies and had to make do with a severely reduced promotions staff after the October layoffs. The BOS also focused on expanding markets for the sale of various component parts, especially bathtubs and porcelain-enameled roofing tiles. The potential for component sales had largely been ignored in the midst of larger difficulties, but the company believed that it might be able to generate sufficient profits from component sales to maintain liquidity while it addressed core issues.

The extent of Lustron's problems remained difficult to quantify. The company had taken a number of positive steps, especially expanding its product line and focusing on profitable dealerships; it remained a viable concern with strategies for overcoming longstanding barriers to the success of factory-made housing. Carl Koch's vision for a "second generation" of Lustron houses also promised greater manufacturing efficiency and greater flexibility for the company, its dealers, and its customers. However, increasingly negative appraisals of Lustron's progress in the national press, a result of the congressional hearings on RFC loan practices, renewed fears in the company that the RFC might "cut its losses" and abandon Lustron to receivership. The company had come a long way in a relatively short time. Now, it needed additional time to demonstrate that its vision was fundamentally sound.

CHAPTER FIVE

A Dream Denied

As the homes, so the state.
—A. Bronson Alcott, *Tablets*

Once a cornerstone of the government's plan to ease the postwar housing shortage, during the summer of 1949 Lustron struggled to maintain its financial viability. On August 30, 1949, the RFC approved a seventh and final loan. The $2 million loan brought the company's total indebtedness to $37.5 million, an amount representing nearly 10 percent of the total outstanding loans in 1949 made by the RFC to industrial and commercial enterprises.[1] Like the three previous loans, it provided operating capital to assist a final push toward commercial production. The ever optimistic Strandlund told reporters that if Lustron could survive the winter of 1949–1950, it would have the opportunity to refine dealer relations, introduce a second generation of houses based on Carl Koch's designs, and address weaknesses in the production system. Privately, however, Strandlund realized he was sitting on top of a powder keg. With operating losses exceeding $1 million for August, the company could not maintain the principal payment on the original $15.5 million loan. The RFC agreed to accept only the interest payment with the understanding that all the loans would be reconfigured once a final assessment of the company's capital requirements could be calculated.[2]

Lustron's continuing dependence on RFC loans underscored its vulnerability to the political controversy swirling around the agency. The company's association with the RFC had been controversial from the beginning. Each time the Banking and Currency Committees of the House and the Senate reviewed RFC lending practices, the Lustron loans received prominent attention. Congressional scrutiny and criticism touched off a cascade of negative publicity that frequently cast the venture as a government-sponsored boondoggle. Although the RFC survived another round of investigations during the summer of 1949, its board felt tremendous pressure to justify the agency's existence and smooth over controversies regarding the administration of its loans.

Congress also questioned a recurring pattern of RFC personnel accepting employment with loan recipients.[3] Lustron had welcomed a number of former government employees into its executive ranks, including several from the RFC. Given Lustron's entanglements with the federal government, Strandlund saw these men as valuable additions to the firm and relied on them to help guide the company's interactions with key federal agencies. He also recognized the danger of such associations. However, he found it increasingly difficult to oppose personnel suggestions from the RFC, especially as the company's financial situation deteriorated. To his dismay, he found the RFC all too willing to attempt to control the selection of key personnel. As it became apparent that a faction within the RFC was targeting Strandlund personally, he began to see his dreams fade slowly away.

A Trojan Horse?

The continuing furor over the Lustron loans focused congressional concerns on potentially corrupt practices within the RFC. The Senate Banking and Currency Committee had expressed a desire since 1948 to implement a general reorganization of the RFC. The committee's interest stemmed from a change in policy in 1947 that separated the RFC from the supervision of the office of the Federal Loan Administrator and allowed the RFC Board of Directors to make loans without formal oversight.[4] The reorganization plan proposed vesting the management of the RFC in a single governor who would replace the five-member board. The committee observed that "the impairment of [the RFC] is the result of deterioration which has occurred in the Corporation's top management structure . . . only a drastic action can restore the integrity of RFC."[5]

Personality also played an important role in the ongoing controversy over the administration of the RFC. J. William Fulbright (D-AK) assumed the chairmanship of the Senate Banking and Currency Committee after the 1948 elections and led the charge for reform. President Truman had a history of personal conflict with Fulbright. On the eve of the Republican sweep of the 1946 congressional elections, Fulbright told a reporter that Truman should appoint a Republican secretary of state, who would be next in line for the presidency since the office of vice president was vacant, and then resign. Truman scoffed at the suggestion, privately called Fulbright "an overeducated Oxford SOB," and thereafter frequently questioned the judgment of "Senator Halfbright." Undoubtedly Fulbright knew of Truman's opinions.[6]

As congressional investigation of the RFC proceeded, the Lustron loans received comprehensive scrutiny. Testifying at Banking and Currency Committee hearings in both houses, Strandlund presented the company's view of its recent relationship with the RFC. He described a troubling chain of events that indicated the RFC had lost faith in Lustron by August 1949 and claimed that events exposed "a conspiracy to take over and operate Lustron" with the intent of destroying "one of the most promising enterprises marking our postwar economy."[7]

In Strandlund's opinion, the origin of Lustron's problems with the RFC dated back to June 1948 when director Harvey Gunderson asked him to attend a meeting in the corporation's Chicago offices. During the meeting Gunderson urged Strandlund to employ E. Merl Young, a loan examiner with the RFC, as assistant secretary of Lustron at an annual salary of $12,000. Strandlund agreed and installed Young in a Washington, D.C., office to serve as a lobbyist for the company. Considering Young's background and connections, it seemed a logical assignment.[8]

A native of Missouri, Merl Young entered government service in 1940 as an assistant messenger for the General Accounting Office. His wife, Lauretta, found employment as a stenographer in the office of the junior senator from Missouri—Harry Truman. After serving in the Marine Corps from June 1942 to October 1945, Young landed a job at the RFC as an "expediter" in the Surplus Disposal Branch. His wife followed Truman to the White House and served as an assistant to Rose Conway, a senior secretary.[9]

Young soon became a close friend of Donald S. Dawson, RFC personnel director. Dawson and Lauretta Young had both been born in Cedar County, Missouri, and their families had a passing familiarity. In 1946 Merl Young won a promotion to the position of RFC examiner and in the following year became a senior financial examiner. In this capacity Young reviewed loan applications and advised the RFC Board on the advisability of acceptance or denial. Described as "brash but likable," Young was extremely ambitious and often "talked up" his connections to the White House. He also intimated that he was related to the president because Truman's mother's maiden name was Young.[10]

Young first crossed Lustron's path during the application and review process for the company's second loan. Apparently he had "concurred" with the loan request when asked by Gunderson to review the Lustron file. The motives for Young's move to Lustron remain unclear. Was he to serve as an RFC insider and report on Lustron's management team? Did his appointment simply reflect the classic Washington pattern of hiring former government officials as lobbyists? Or was it purely a patronage position? Strandlund's ready assent in hiring Young also calls into question his motives. Records do not reveal that he was upset by Gunderson's meddling at the time, and Strandlund probably saw the value of hiring such a "well connected fellow" as Lustron's man in Washington. Further, Strandlund was undoubtedly reluctant to deny the wishes of an RFC director "whose integrity [he] never had reason to question" since he knew the company would require additional loans.[11]

However, Strandlund's relationship with Young began to deteriorate soon after he joined the company. With Young in Washington and Strandlund in Columbus, Young remained free from close managerial oversight. Apparently, Young believed that his duties as assistant secretary of Lustron included working for Truman's 1948 reelection campaign. A publicity pamphlet published by the Democratic National Committee listed him as a "campaign assistant."[12] Young also made a series of "political trips" to Texas, Illinois, Ohio, Michigan, New York, and Missouri during the campaign, and he joined Truman's whistle-stop entourage during the week before the election.[13]

Soon after Truman's victory, Strandlund received a call from Harley Hise, chairman of the RFC, requesting that he join Hise and Young at the Statler Hotel in Washington. At the meeting Hise suggested that Strandlund promote Young to vice president at an annual salary of $18,000. Strandlund reported that Hise wanted to see Young "get something out of it," meaning Young's efforts on the campaign trail. Again Strandlund assented, and Young's promotion became effective in December 1948.[14]

Another facet to the Young saga involved his association with the F. L. Jacobs Company, an appliance and vending machine manufacturer with several outstanding RFC loans. Rex Jacobs, president of the company, hired Young in December 1949 at $10,000 per year "to promote the installation of vending machines in the Washington area." Subsequent testimony revealed that neither Strandlund nor RFC officials knew about Young's "double-dipping." Strandlund claimed that he "would have taken measures to correct the situation if [he] had known" about the arrangement.[15]

As a self-described "man with pretty close contact in high circles," Young may have also played a role in the selection of Walter Dunham, a Republican, to the RFC Board in February 1949. The RFC Act required that no more than three of the five board members belong to the same political party. With three Democrats already in place, administration officials naturally sought a liberal Republican, and Dunham seemed a logical choice. Interestingly, Dunham's closest White House contact during the interview process was Donald Dawson, the former RFC personnel director, then serving as a personnel assistant to the president. When Dawson asked Dunham if he "could work in harmony with the Democratic Party," Dunham replied that he "could work in harmony with anybody."[16] After Dunham's appointment, Young commented to Strandlund that "his man" had been selected and that Strandlund should come to Washington as soon as possible to meet the new director.[17]

When Strandlund, Young, and Dunham met in March 1949, Dunham informed Strandlund that he expected "to take a great deal of interest" in Lustron. Dunham further opined that the RFC had not treated Strandlund fairly and that "if [he] had been handling the case [he] would have given Lustron $50,000,000 and let the project go ahead without interference." Strandlund then invited Dunham to the plant for an "informational" tour, and on May 4 Dunham and Richard Dyas, chief of the

RFC Housing Branch, came to Columbus and told Strandlund that they were impressed with the operation and pleased with the company's progress.[18]

Within a month, however, Dunham's assessment had apparently changed. When he visited the plant on June 22, Strandlund reported that Dunham "had done a complete about face and reversed his previous appraisal" and "urged that practically all of the top executives in Lustron be fired." In deference to Dunham's suggestion, Strandlund fired sales vice president Richard Jones and agreed to hire V. A. McKechnie, a business associate of Dunham's, as Lustron's vice president of operations. "I learned thereafter that [McKechnie] was in almost continuous secret communication with Mr. Dunham," Strandlund later told the Fulbright committee.[19]

During one of Strandlund's frequent visits to Washington to testify before Congress, he met with Dunham and Young, who informed him that Rex Jacobs would visit the Lustron plant in July. On July 6 Jacobs came to Columbus. While he was in Strandlund's office, Dunham called. Dunham asked if Jacobs was there and urged Strandlund to "talk freely" with him. After a plant tour, Jacobs offered a proposition. If Lustron would use the Jacobs Company's washing machines in its houses, he would offer a $15 kickback per machine to Merl Young. Strandlund explained that Lustron had an exclusive contract with the Thor Corporation for the combination dishwasher/washing machine and that "the house could not stand the $15 extra charge for Young." Jacobs asked, "what shall I tell Walter [Dunham]?" Strandlund replied, "tell Walter to keep his shirt on." Recalling the incident several years later, Strandlund observed, "From then on I was in trouble."[20]

Whether a cabal of RFC officials, with the help of Lustron "insiders" Young and McKechnie, attempted to engineer a takeover of Lustron remains unclear. Certainly this would have been the lesser of two evils for the RFC. Given the political climate, the agency could ill afford to see Lustron fail without evidence of "proactive involvement." However, the limits of proactive involvement held dangers as well, for such interference would likely be cast as undue influence by congressional critics of the RFC. On the other hand, if a takeover succeeded and Strandlund's management team was ousted, the RFC could claim that it was protecting the interests of the taxpayers by reorganizing the company. What is clear, however, is that Dunham, Jacobs, Young, and other representatives of the RFC began in earnest to pressure Strandlund to relinquish managerial control beginning in August 1949.

Boxing Carl

On August 15, as negotiations for the final $2 million loan proceeded, Strandlund again met with Dunham. Strandlund reported that Dunham "stated in threatening language that [the RFC] would close down the plant, stop the operation, [and] consolidate it with some other enterprise, the identity of which he would not disclose," unless Lustron's board immediately voted to transfer full control of the company to a management committee appointed by the RFC. Stunned, Strandlund met with RFC director Harvey Gunderson, who informed him that "no such action would be taken."[21] However, as a condition of the final loan agreement, the RFC insisted on a clause requiring Lustron to accept "a committee nominated by RFC with full powers of investigation, to confer and advise with borrower's management and report direct to RFC." On the same day, the RFC Board passed a resolution prohibiting Lustron from purchasing additional inventory without the specific approval of the agency."[22]

Strandlund's furor over the attempt to "strangle Lustron's operations" increased on August 31 when the RFC appointed Rex Jacobs and Edward J. Hunt, a former RFC employee then with the Chrysler Corporation, to investigate Lustron's progress and prospects. On September 7 and 8, Jacobs, Hunt, Young, and Dyas inspected the plant and began compiling data for submission to the RFC Board. Interestingly, the Jacobs-Hunt report, as it became known, stated that Lustron's operations were "sound" but recommended a wholesale change in the management structure with promotions for Young, to president, and McKechnie, to executive vice president.[23] Apparently RFC chairman Hise agreed and telephoned Strandlund on September 14 requesting the resignation of all Lustron's officers and directors except Strandlund himself. When Strandlund asked to see the Jacobs-Hunt report, Hise suggested that he come to Washington to discuss the matter before the entire RFC Board.[24]

Strandlund arrived in Washington on September 26 and found only Dunham and fellow director William E. Willett in attendance. Dunham reiterated the request for the resignations and also informed Strandlund that he would be required to relinquish control of 60,000 of his 86,000 shares of Lustron stock without compensation. The shares would then be distributed to the new, and as yet unnamed, management team. Again Strandlund approached Gunderson after the meeting. Gunderson assured him that he could retain his stock but offered no hope of forestalling a general reorganization. On October 4 Strandlund received a telegram from Hise informing him that the "resignations must be in our hands on next Monday, October 10, or it will be necessary for this Corporation to take immediate action designed to protect its interests."[25]

On October 5 Strandlund wrote Hise a lengthy letter requesting a copy of the Jacobs-Hunt report and suggesting that the agency delay any action until the engineering consulting firm Booz, Allen & Hamilton completed its analysis of Lustron's status and prospects.[26] Due for completion in December, Strandlund promised to implement the suggestions contained in the Booz, Allen report. The next day Strandlund received a call from Merl Young "in which he strongly urged [him] to accede to Mr.

Dunham's demands and stated that [he] would have nothing to fear from them." Young apparently assured Strandlund that he could remain as chairman of the reorganized firm with a lifetime contract and a salary increase. Strandlund reported that he "could not find it in [his] conscience to accept such a proposal."[27]

As Strandlund and the RFC battled behind the scenes, the Lustron situation drew increasing attention in the House of Representatives as a number of Republican members took the opportunity to implicate the Truman administration in a scandal. On October 11 Representative Albert Cole introduced a resolution authorizing an investigation of the RFC and Lustron.[28] His accompanying remarks offered a rationale for the inquiry:

> The following questions should be answered. By what authority were these loans made? Can the experiment succeed? If it succeeds, will a private monopoly of the prefabricated housing industry be created? If it fails, will the Government take over and socialize the prefabricated home building?[29]

Representative Charles Vursell (R-IL) rose in support of the resolution and offered a detailed account of Lustron's history that focused on the origins of White House support, the lack of invested private capital, and the high cost of the houses to consumers. The only positive aspect of the entire episode, Vursell sarcastically noted, was that Wilson Wyatt "resigned, and at least that much was accomplished for the good of housing in the Nation." Vursell also reached for an irresistible comparison of the Lustron venture with Soviet central planning when he stated that "these houses should go over big and be a luxury in Russia where the Government tells you where to live and in what type of house you shall live in."[30]

Lustron again took center stage during debates on the floor of the House in early October over proposed amendments to Section 102 of the National Housing Act. One of the amendments included a provision authorizing $25,000,000 for marketing and financing assistance for dealers of prefabricated houses. The debate highlighted a frequent dilemma for critics of the company and evidenced the separation of "dealer interests" versus "company interests" in discussions about the future of Lustron. Many congressmen who found it easy to criticize the company became ardent defenders of the company's dealers, who represented the archetypal virtuous small businessmen. Could the government so easily abandon them?

Representatives Cole, Jesse Wolcott (R-MI), Brent Spence (D-KY), and A. S. Mike Monroney (D-OK) engaged in a lively exchange in Congress and in the press during October regarding the efficacy of continued government support for prefabricated housing companies and their dealers. The debate illustrated various perceptions of the Lustron situation. Cole and Wolcott favored the immediate liquidation of the company and the initiation of legal action against Strandlund. "Over $3,000,000 worth of houses are crated and stand forlornly on trailers at the Lustron plant, with no takers," Cole said, and "the problem of sales and distribution of these houses is still . . . a mystery."[31] Both also argued that the $25 million on the table for prefab dealers represented another "subsidy" to Lustron that would effectively shut out dealers of "more worthy" prefabricated housing firms. Spence and Monroney countered that at least twelve manufacturers of prefabricated housing would benefit from additional funding for dealer financing and marketing programs. The money would "move the product, *not* aid in the manufacture of more product" and "put prefab dealers on a par with conventional contractors, who have access to interim financing," Spence argued.[32]

Regarding Lustron, Monroney asserted that the government needed to find out whether marketing and dealer problems could be solved. "Like it or not, it has $37.5 million in government money," he said, and apparently "cannot reach the market on account of lack of some type of financing which we are freely giving to the conventional builders all over the country." Why squander the government's investment without addressing dealer issues? If more funding will solve the problem, then Lustron could repay its debts. If not, at least the government retained its right to "salvage a portion of this investment." Ultimately, Congress rejected Spence and Monroney's arguments and withdrew appropriations for additional support of prefab manufacturers and their dealers.[33]

Meanwhile, an RFC report citing Lustron's inability to make principal payments on any of its loans swayed many congressmen to consider "closing the book" on the company.[34] In October, Lustron laid off over 2,000 employees, leaving barely 1,000 remaining from a high of 3,400 in June. Worse yet, the company owed over $20 million to the RFC in principal and interest payments that came due in January 1950. Lustron's relationships with American Community Builders and the Galbreath Mortgage Company also came under suspicion when congressional investigators revealed that both firms held outstanding loans from the RFC. Galbreath denied drawing on the loan, and American Community denied that the intent of its loan application was to finance the purchase of 2,000 Lustron homes for the Park Forest project.[35]

Back in Columbus, Strandlund scrambled to regain control of his enterprise. His attention turned first to Merl Young. "I concluded that there was no doubt as to Merl Young's disloyalty to our organization," he said, but removing this "Trojan Horse" was fraught with danger. Rather than fire Young outright, Strandlund chose to relocate him to Columbus and assign him "duties in which he would be more useful to the company." Strandlund informed Young of this by letter on October 10, and Young responded with a letter of resignation. Young related his regrets but indicated that he felt compelled to resign due to the company's "failure to adopt policies of retrenchment" that he had "urged for some time past." This was news to Strandlund, who claimed Young had never discussed Lustron's "retrenchment" with him. Curiously, Young's letter set his resignation date at November 30, nearly six weeks hence. "It occurred to me that he may have felt that by November 30 there might be such changes in management as would permit him to withdraw his resignation before its effective date," Strandlund surmised.[36]

Satisfied for the moment with the Young situation, Strandlund prepared a strategy to forestall what he believed was an RFC "takeover" of Lustron. The proposal, completed on October 24, called for a new board of seven directors, three selected by the RFC, three selected by shareholders, and the seventh chosen by the previous six. Further, Strandlund agreed to transfer control of his shares to a "voting trust" and included a provision giving the RFC the right to convert the amount of Lustron's indebtedness into common stock with full voting rights if the company failed to meet a revised repayment schedule.[37]

RFC chairman Hise wrote to Strandlund on November 9 advising him that the RFC directors had considered the proposals and "did not find them satisfactory."[38] On December 18, Richard Dyas, assigned by the RFC Board as liaison for Lustron, asked the company to submit a more detailed reorganization proposal no later than January 6, 1950. If the company failed to devise an acceptable plan, the RFC would "give immediate consideration to the protection of the Government's interests by other means at its disposal."[39] Clearly this meant foreclosure, and Dyas offered to come to Columbus and assist in the formulation of a revised proposal.[40]

Lustron submitted the revision to the RFC via messenger at 3 P.M. on January 6, 1950. A reorganized board headed by Phillip Klutznick, former FHA commissioner and president of American Community Builders, led the list of changes. Additional members of the board included Harley Hise; current member Howard E. Buhse, a partner in Hornblower and Weeks; current member Louis Leverone, president of the Automatic Canteen Company; Strandlund; and two members to be selected later, "preferably outstanding figures in the mortgage, building, or construction fields." Next, the proposal called for the creation of the Lustron Development Corporation, a subsidiary headed by Klutznick, which would give "particular attention to distribution problems and to the promotion of rental and other projects utilizing Lustron homes." Klutznick would receive 25 percent of Lustron stock, partly from Strandlund's personal holdings. Strandlund, however, would remain as president of the Lustron Corporation.[41]

Perhaps the most significant aspect of the proposal was a request for an additional $3–5 million in working capital, "to be disbursed as needed." The amount reflected Strandlund's ever optimistic prediction of Lustron's output during the first six months of 1950. He estimated that the company could produce—and, more importantly, sell—2,800 houses, a conservative assessment in his opinion, considering "the very real probabilities of project sales well in excess of 5,000 houses." To support his claim he submitted a list of twenty fleet sales projects in various stages of development, eight of which were for military housing.[42] The proposal also requested the rescheduling of all of Lustron's outstanding debt to the RFC to mature in 1958 at 4 percent interest payable quarterly, "but only to the extent that net income provides funds for that purpose." In other words, until Lustron returned a profit, it would not have to pay down its debt. Further, the plan provided for the removal of restrictions attached to the stock owned by Strandlund and his wife and for the elimination of the personal guarantee placed on his assets, both requirements of the original $15.5 million loan.[43]

In Strandlund's view these requests and the appeal for additional capital represented valid and necessary aspects of the reorganization plan. As he told Dyas, "the likelihood of securing any substantial additional [private] equity capital for the company at this time appears extremely remote."[44] Indeed, on January 11 Strandlund received a telegram from the RFC advising the company that the directors had "considered and declined" the proposal.[45] In rendering its decision, the RFC relied heavily on the completed report from the consulting firm Booz, Allen & Hamilton, which offered few reasons to risk additional investment. Distilled to its most salient point, the report stated that "Lustron . . . has produced a house which is in a price class in which there is no real opportunity for large volume; yet the only hope for a profitable enterprise, against this large investment, is a large volume."[46]

Feeding Frenzy

With appropriations for aid to prefabricated housing manufacturers and dealers eliminated by Congress and with an unfavorable report on Lustron from a respected consulting firm, the RFC moved quickly toward foreclosure. *Newsweek* quoted an unnamed RFC official who observed that Lustron's proposed reorganization "wasn't a reorganization plan at all."[47] Since the company had defaulted by not keeping its loan payments current, the RFC could technically foreclose at any time by calling the loans. The only barrier to immediate foreclosure was the desire of the Senate Banking and Currency Committee to hold additional hearings to "get more light on what [was] going on between Lustron and the RFC." Chairman Hise agreed that further inquiry by the Fulbright committee would bolster the RFC's decision to end the Lustron experiment.[48]

The delay, however, also renewed Strandlund's hopes of working out a new deal, especially after an RFC official told the *Columbus Dispatch* if Strandlund walked in with an acceptable reorganization plan "we would not arbitrarily tell him he is just too late." But the company evidently squandered its "grace period" by offering no additional plans to the RFC. Apparently Strandlund refused to compromise since he believed he was being "railroaded" by the agency.[49] Further, he remained confident in his ability to convince the Fulbright committee to halt the foreclosure. Yet after only three hours of testimony in executive session, the committee essentially washed its hands of the matter and "quickly turned it back to the RFC." Fulbright said he "saw no reason to interfere" and believed the committee was "not in a position to override the RFC."[50] Lustron's old friend Senator Ralph Flanders stood as the lone voice on the committee in defense of the firm. Flanders called the Lustron house "the best mass produced design for housing" and said that the company had merely "run into difficulties which beset all new businesses."[51]

The specter of foreclosure set into motion a flurry of activity among Lustron's creditors, dealers, and employees. Columbus banks took the precautionary measure of requiring the presentation of all Lustron checks at the bank on which they were drawn. This was an effort to prevent losses on uncleared checks, since in a foreclosure all corporate accounts would be immediately frozen.[52] And yet, after the RFC rejected the January 6 reorganization plan, many dealers traveled to the plant with the intention of purchasing as many houses as possible. W. W. Lobdell, president of Lobdell Realty Company, the Lustron dealer in Rockford, Illinois, came to buy "all the houses Lustron will sell" to complete a 200-house Lustron subdivision. Other dealers also vied for remaining inventory. Quickly, the houses Representative Cole had said stood "forlornly on trailers at the Lustron plant, with no takers," were gone. The company also received over 6,000 additional "firm" orders, defined as orders accompanied by a 10-percent cash down payment.[53]

Lustron employees agreed to keep "at the job" until the RFC ordered the company to halt production. Albert Henry and Betty Lintner of the Order Department reported that they were "busier than ever" and expressed confidence that the company could succeed if given more time and more money. "I don't think a few more dollars would hurt," Lintner said; "I think the company has shown enough progress that it is ready to go over the top." Paul Valentine, president of the carpenters' union local at the plant, pledged the support of the AFL to "stop RFC action and obtain more money for Lustron."[54]

Despite optimistic outlooks and promises of action on the part of Lustron supporters, the only major decision left to be made was when and how, not if, the foreclosure suit would proceed. On January 17 the RFC Board ordered its legal department to prepare for court action and asked it to consider whether the next step should be "foreclosure, receivership, or some other legal action to seize the company's assets." A simple foreclosure and sale of assets would have been the easiest course of action, but as Strandlund pointed out in the January 6 plan, it "would probably recover only five cents on the dollar."[55] RFC attorneys suggested that a better course of action would be to foreclose, appoint a receiver, and explore additional options. Chairman Hise, consulting with Defense Department officials, suggested that the plant could be used to produce badly needed military housing, since Lustron had an excellent reputation and record in this market. When Representative Cole got wind of the idea, however, he strongly objected at yet another attempt "to put the Government into the prefabricated housing business."[56]

Finally, on February 21, the RFC issued a press release announcing the foreclosure action: "We have withheld action for several weeks . . . nothing concrete has resulted . . . we can wait no longer." The agency filed the suit in the court of U.S. Federal District Judge Mell G. Underwood. It claimed $22 million from Lustron and $15.5 million from Strandlund since he had given his personal endorsement as part of the original loan agreement. The suit also asked for the appointment of a receiver to oversee the disposal of Lustron's assets. Depending on circumstances, though, a receiver could also choose to continue operations. During the foreclosure hearings, RFC lead attorney Frank Harrison argued for terminating operations. Already in default, Lustron was "in jeopardy of further loss at the rate of $35,000 a day," he said. John C. Harlor, Lustron's attorney, protested that the company was "at peak production and would be operating profitably within a year."[57]

Lustron's dealers also emerged as important players in the foreclosure drama. Richard L. Krabach, the dealer in Lima, Ohio; Arthur Padula, the New Jersey dealer; and W. W. Lobdell had organized the Lustron Dealers Association in mid-1949. By February 1950 the Association represented 221 dealers. The original intent of the Association had been to increase communication among dealers and to serve as a clearinghouse for useful marketing and sales information. However, as Congress stepped up its investigations of the RFC-Lustron affair, the Association became more aggressive and politically savvy in drawing attention to the plight of Lustron dealers. The Association frequently contacted members of the Fulbright committee and other government officials to inform them of the dealer perspective on the Lustron saga.[58]

After the RFC filed the foreclosure suit, the Association asked the Fulbright committee to convene hearings examining the implications of foreclosure and possible liquidation on the dealer network. On March 1 dealers and other Lustron supporters offered a number of arguments for "the maintenance of Lustron as a going concern." Lobdell opened the hearings with the firm assertion that the majority of Lustron's dealers believed in the company. He also claimed that if a receiver would permit continued production, dealers had customers waiting to purchase the houses. "We are right on the verge of really going to town," said Lobdell.[59] He told the committee that Lustron's dealers did not wish to interfere with the foreclosure suit but desired to publicize their interests and general optimism regarding the future of the corporation. After all, he argued, none of the dealers had ever been called to testify before a congressional committee investigating Lustron or the RFC. "We don't care who the management is," said Lobdell; "we are ready to go ahead and build the number of houses necessary to put Lustron back on its feet."[60]

Several committee members questioned him about marketing difficulties. Lobdell admitted that his dealership had faced many challenges; excessive site assembly time of the first houses, problems with FHA guarantees, and building code restrictions, but he maintained that nearly all of the problems had been solved. With time, dealers could overcome every barrier to success. "People just didn't believe it could be as good as it looked . . . because it is so radically different," Lobdell stated. "Now FHA knows it is a good house . . . we have had wonderful cooperation [and] our sales are pyramiding."[61]

Lobdell also explained that Lustron's dealers had invested over $10 million in the enterprise and that liquidation of the company would cause severe financial harm to all but the most profitable dealers who had already paid off their debts. "If I thought for 1 minute that this house didn't have a future, we would have quit a long time ago," he told Fulbright. Lobdell then produced telegrams from ninety-seven dealers who had firm commitments for over 8,000 houses during 1950. Almost every dealer attached glowing comments from their customers about the Lustron house, and most blamed "stupid conflict" between the RFC and Lustron for negative perceptions in the press. "Ask the homeowners," Lobdell urged.[62]

He then related an incident illustrating consumer loyalty that occurred during a meeting between Lobdell and a consultant from Booz, Allen & Hamilton who had been in Rockford to research marketing issues for the RFC report. The consultant refused to believe Lobdell when he claimed that customers loved the houses. The consultant said the customers would "talk them up, regardless, just to protect themselves." Lobdell then put his telephone on speaker and began calling his customers: "I offered those people $2,000 more than what they paid for their house, and without exception everyone we called said: 'We will not give this house up unless you can get us another Lustron house to move into.'"[63]

Lobdell also suggested that Lustron's future could be guaranteed by continued expansion of military and rental project sales. He offered the committee evidence of the success of the Quantico project and the cost benefits of using Lustron houses rather than conventionally built houses in remote locations. The suitability of Lustron houses for rental projects also justified continued production, Lobdell maintained. "You don't care how dirty the people are that live in the house. All you do is take a hose and wash it off," he said.[64]

R. L. Krabach and Arthur Padula echoed Lobdell's sentiments and reported much the same experiences in their territories. Padula was especially critical of the RFC's rationale for foreclosure, charging that the RFC did not want to "get control of it," but desired to destroy the company and close an embarrassing chapter in its history. Leo Goodman, director of the national housing committee of the Congress of Industrial Organizations (CIO), agreed and ended the hearings with an impassioned defense of the entire prefabricated housing industry. "We are seeking to urge the Congress to take steps to help establish an industry which, in the present complex production systems, cannot get off the ground by itself," said Goodman. He based his argument primarily on the potential contribution of the industry to national defense. Firms such as Lustron, he claimed, would be able to produce a "large volume of homes, easily assembled, in a period when the possibility of attack is a real one."[65]

Goodman also compared the government's "short-sighted approach" to prefabrication with the problems faced by early automobile manufacturers, who also found it difficult to deal with people "who failed to realize the tremendous financial potential." Even if Lustron only served the needs of the military housing market, it would serve a vital national concern. Goodman concluded his testimony by charging that the RFC "was out to kill this industry." When Fulbright asked for specifics, he raised the issue of "suicide loans"—the sixty-day term loans that gave Lustron little flexibility for repayment and whose default formed the basis of the foreclosure suit.[66]

Apparently, however, the hearings provided Lustron's dealers little benefit beyond the satisfaction that at least someone had listened seriously to their concerns. In Columbus on March 6, Judge Underwood granted the foreclosure decree to the RFC and appointed Clyde M. Foraker, chief of the Columbus field division of the Internal Revenue Service, as receiver "for a trial period of thirty days." Judge Underwood also delayed any liquidation sale of Lustron's physical assets for at least forty-five days and ordered the RFC to work with the company on alternatives to liquidation.[67] Foraker immediately fired most of Lustron's top management, including Strandlund. Only D. W. Boylan, corporate secretary and comptroller, and W. A. Matheson, general sales manager, survived the foreclosure.[68]

Judge Underwood's decision and Foraker's house cleaning brought the main chapter of Lustron's history to a close. The RFC gained control of the company and guaranteed that the agency would receive at least a portion of its investment. Even Lustron's supporters agreed that Strandlund could not be a part of the company's future, if indeed the firm had a future. Although the foreclosure decree ended Strandlund's personal involvement in Lustron, he remained personally liable for $15.5 million of corporate debt. He felt betrayed by the RFC and believed that he was a victim of a conspiracy to destroy his company.

The March 6 foreclosure decree, the appointment of Clyde M. Foraker as receiver, and the dismissal of Carl Strandlund and all but two of his management team ushered in the final chapter of the Lustron experiment. Instead of going gently into that good night, however, the company remained at the center of ongoing political and legal controversies. The taint of scandal also trailed the firm far beyond its commercial demise. With $37.5 million of government funds at risk, officious recriminations reverberated throughout Washington as the "blame game" swung into full force.

Requiem for a Prefabricator

Reactions to the plight of the moribund firm ran the gamut from glee, to anger, to despair. Other prefabricated housing companies wondered why Lustron had received a majority of federal funds allocated to aid the prefabricated housing industry. Congressional critics attacked the RFC for a lack of oversight, accused Lustron's management of incompetence, and questioned the intent of federal support for prefabrication in general and Lustron in particular. Strandlund and his colleagues maintained that Lustron had been the target of a scandalous takeover attempt by a faction of RFC and Truman administration officials. Others believed that Lustron had been sabotaged right at the moment when it had overcome the inevitable birth pangs of a new enterprise.

Senior engineer W. V. Burnell of Stone & Webster, the author of several analytical reports on Lustron's business, believed that the company had been "deliberately scuttled" to protect the tarnished reputation of the RFC. "Why in Hell can't *somebody* in public life come out and say so," he wrote to a former Lustron vice president.[69] Whether additional time or capital would have produced a different outcome remains unclear. But the reality of foreclosure, bankruptcy, and liquidation initiated a variety of attempts to salvage something of value from the largest experiment in prefabricated housing in the history of the United States.

After accepting his appointment as receiver, Clyde M. Foraker began a search for companies or investor groups willing to purchase Lustron as a going concern. The RFC favored this approach because it offered the best way for the agency to recover its total investment. Reflecting the RFC's desires as well as his own concerns for the best interests of the taxpayers, Judge Underwood on April 3, 1950, granted an indefinite extension of the thirty-day "trial period" of Foraker's receivership. Judge Underwood realized that Lustron would be more attractive to potential investors while in operation, with numerous orders on the books from dealers who feared immediate liquidation.[70]

Several possibilities soon emerged. The prefabricated housing firm Gunnison Homes contacted Foraker but quickly withdrew from consideration after its board rejected the purchase as excessively risky. Fruehauf Corporation, manufacturer of Lustron's delivery trailers through its Commercial Home Equipment subsidiary, also considered forming an investor syndicate, primarily to protect its $3 million investment in Lustron. However, chairman Roy Fruehauf ultimately decided to write off potential losses and keep his firm out of the economic and political vortex. Finally, Phillip Klutznick, president of American Community Builders, approached Foraker with a plan to purchase Lustron. Klutznick had been slated to become chairman of Lustron had the RFC approved the January 6 reorganization plan. Klutznick's syndicate included other merchant builders who wanted to use Lustron houses for subdivision development. However, when the press reported their plans, the Klutznick group withdrew its offer. The combination of financial risk and potential association with the Lustron/RFC imbroglio effectively doused the enthusiasm of each of the company's potential "white knights."[71]

The failure to attract investors increased pressure on Foraker and Judge Underwood. The RFC urged the judge to approve an auction of Lustron's physical assets to facilitate the agency's purchase of the plant and its disposal on terms favorable to their interests. Underwood agreed, and on June 6 the RFC, the only bidder, acquired the Columbus plant for $6 million at an auction supervised by the court. RFC officials also announced their intention to continue the search for private investors. At a separate auction on the same day, the Lafayette Steel Company of Detroit, Michigan, outbid several other companies and purchased $1.5 million of high-grade steel in Lustron's inventory for $645,000.[72]

Yet on the evening of June 9 events took a startling turn. Three Lustron dealers from Illinois went to the Chicago home of U.S. Federal District Judge Phillip L. Sullivan and requested an injunction to block the sale of Lustron's plant to the RFC. The dealers claimed that Lustron owed them a total of $4,842 and feared the sale would preclude collection. Judge Sullivan issued a temporary injunction that ordered Foraker to delay the transfer of deeds of ownership to the RFC.[73] Michael Gesas, an attorney representing the dealers, also requested that Judge Sullivan issue an involuntary bankruptcy motion. Such a motion would place all of Lustron's creditors on an equal footing rather than permit the RFC to assume the position of a preferred creditor to the detriment of all others. Gesas claimed that the Chicago district federal court had superseding jurisdiction because Lustron was incorporated in Illinois. Further, Gesas argued that a bankruptcy motion took precedence over a receivership, thus invalidating the actions of Judge Underwood in Columbus. Joseph H. Schwartz, one of Lustron's attorneys, supported Gesas and filed a consent motion informing Judge Sullivan that Lustron's officers and directors desired involuntary bankruptcy. Corporate secretary D. W. Boylan, resentful of the RFC, believed that an involuntary bankruptcy initiated by Judge Sullivan and overseen by an Illinois bankruptcy court would provide the best protection for Lustron's dealers.[74]

The emerging jurisdictional conflict left the company in limbo pending an appeal to the U.S. Circuit Court of Appeals or a negotiated settlement between the parties. Such a settlement seemed unlikely when neither judge indicated a willingness to retreat from his position. On June 14, Judge Underwood, ignoring Judge Sullivan's temporary injunction, approved the sale of the plant to the RFC and the sale of the steel to Lafayette. On the same morning, Judge Sullivan ordered Lustron into involuntary bankruptcy and issued a permanent injunction barring the sale of Lustron's assets. However, U.S. Marshall Harold Claypool, acting on Judge Underwood's orders, formally transferred deeds of ownership to the RFC at 11 A.M. Judge Sullivan issued his orders at 11:20 A.M.[75]

Judge Underwood was furious. That afternoon he announced that unless the Department of Justice began an investigation of the entire Lustron affair, he would call a grand jury "to do the job." Strandlund, following events closely from his Columbus home, issued a statement welcoming further inquiry: "Lustron has been smeared and slandered to the point where I believe it is time the public should know the true facts." RFC director Gunderson responded that the RFC had kept the Department of Justice advised of all potential irregularities in the Lustron files. "I don't know of anything of a criminal nature on the files," he said.[76] Following a conference on June 18 between Justice Department attorneys, U.S. District Attorney Raymond J. O'Donnel, and Judge Underwood, the Department of Justice pledged to investigate Lustron "without unnecessary delay." Judge Underwood agreed to postpone impaneling a grand jury pending the results of an investigation. Justice Department attorney William Delaney then released a statement assuring the public that "all evidence indicating criminal aspects are going to be investigated fully and completely . . . in view of facts developed by Reconstruction Finance Corporation investigators and men associated with the receiver."[77]

Pending an appeal to the Seventh Circuit Appeals Court in Chicago and the results of the investigation, each side could do little but assemble the necessary documentation to support its case.[78] Meanwhile, acting on information uncovered by RFC and Justice Department investigators, the Senate Banking and Currency Committee convened hearings to explore the relationship between Lustron and the Commercial Home Equipment Corporation, a subsidiary of the Fruehauf Corporation that manufactured Lustron's house delivery trailers. The investigation had uncovered financial inconsistencies and a potential conflict of interest between the firms. Testimony revealed that Commercial Home had overbilled Lustron approximately $500,000 for equipment that it never delivered. Paul Cotter, chief of the RFC investigation division, said that under a 1948 contract Commercial Home agreed to rent tractors and trailers to Lustron and had failed to deliver, but not to charge, the company for forty out of 200 units. "The conclusion would appear to be inescapable that there was a fraudulent overbilling," said Cotter.[79] Records indicated that Lustron had paid nearly $3 million in rent for the trailers since 1948. Further, Cotter charged that Lustron officials had "looked the other way" because Paul O. Buckley, a director of Commercial Home, also served on Lustron's board. In view of the interlocking directorate and Buckley's "dual capacity," Cotter asserted that "it is more than a possibility . . . that fraud has been committed against the RFC."[80]

The committee unanimously agreed, and on July 19 issued its report and recommendations to the Senate. The report criticized the RFC and Lustron officials and recommended a wholesale revision of loan oversight procedures. It concluded that the RFC had been "ineffectual in its supervision of the transportation phase of the Lustron operation" and that officials charged with protecting public funds loaned to Lustron "exhibited an inability to detect irregularities and an indifference to unbusinesslike procedures." The committee also criticized Lustron's management for financial incompetence and for allowing Paul Buckley to sit on both boards.[81]

An end to the legal wrangling over Lustron's demise appeared at hand when the Seventh Circuit U.S. Court of Appeals held a special hearing on the Lustron situation in October. Foraker asked the Court to approve back pay for the seventy remaining Lustron employees, who had been working without pay since the conflict began in mid-June.[82] On October 23, the Seventh Circuit Appeals Court rendered its decision in two separate cases covering the RFC plant sale and the sale of steel to Lafayette. The Court upheld Judge Sullivan's injunction against the sale of Lustron's assets and declared that once Judge Sullivan had ordered Lustron into bankruptcy, the Ohio district court "lost all power to proceed in any respect with regard thereto without the consent of the bankruptcy court." Thus the sale of the plant and raw materials were invalid and constituted a violation of the Bankruptcy Act because it placed the RFC in a preeminent position over all other Lustron creditors. The decision effectively terminated the receivership and placed authority to determine the settlement of Lustron's estate within the jurisdiction of receivers appointed by the federal bankruptcy court in Chicago. The RFC and Lafayette Steel announced their intention to appeal the decisions to the Supreme Court.[83]

Meanwhile, a final chance to preserve the company as a going concern emerged in the wake of the Truman administration's desire to put the RFC situation to rest. Fed up with negative publicity and implicitly questioning the future viability of the agency, President Truman had submitted a plan of reorganization to Congress in May 1950. In June Congress passed Reorganization Plan No. 23, which transferred to the Housing and Home Finance Agency (HHFA) all functions of the RFC pertaining to the financing of prefabricated housing.[84] The objective was to place the administration of federal policy relating to the prefabricated housing industry within HHFA, which also held responsibility for the defense housing program. The idea of preserving Lustron to serve the military housing market now began to gain support in Congress.[85]

The advent of the Korean War and the accompanying expansion of defense production again raised the issue of providing adequate housing for the military and for civilians working in defense plants. Lustron houses had already proven their stamina in this market, and procurement officers from each service branch testified to the efficiency and economy of the houses from Quantico to Ft. Dix to remote bases in Alaska. Old enemies now rallied around the idea. Ironically, Strandlund, Representative Cole, and the RFC now found themselves on the same side of the Lustron issue. The RFC favored the idea because it provided an opportunity to recover its investment. Cole and other congressional critics of Lustron admitted that serving the defense market was indeed one area where the company had achieved success. Further, they believed that it would be a "crime" to close down the plant when such an obvious need for military housing existed. Success in the military housing market would also lessen the sting of rationalizing $37.5 million in previous government investment.[86]

Lustron's deposed founder now saw an opportunity to revive and perhaps regain control of his company. In a letter to newly appointed RFC chairman W. Elmer Harber in December 1950, Strandlund reminded the RFC that "the President has pointed out the extreme importance to defense production of a program for prefabricated housing." He also argued that Lustron houses met the diverse needs of the military housing market better that any other house, prefabricated or otherwise. It was a proven, successful product. "Under these circumstances, it is imperative that the Lustron operation and facilities be reactivated immediately," he said. Strandlund then submitted a proposal for a "speedy reactivation" of the plant. Details of the plan included dismissing the foreclosure and bankruptcy suits and paying all unsecured creditors "to clear the slate." Next, he called for the consolidation of all outstanding Lustron loans and requested an additional $3.5 million in government funds to provide working capital during the reactivation phase. Lustron's access to external capital would also be facilitated by restructuring the company's debt, which would remain subordinate to all capital acquired from private investors. This would offer an incentive for private investors, who could be assured that they would not have to "stand in line" behind the government to receive a return on their investment.[87]

Realizing that a request for additional government investment in Lustron would be extremely controversial, Strandlund argued that it would provide the only way for the RFC to recover its entire investment. "Counsel for the RFC has repeatedly stated that [the RFC] did not expect to realize more then Five Million Dollars from the forced liquidation of the Lustron Corp.—a direct loss of $31,500,000 to the government," he wrote. However, a revitalized Lustron could quickly become profitable by serving the military because it would then face none of the vexing marketing problems that inhibited success in the general market. For a "nominal" additional investment of $3.5 million, the government could reap significant benefits. These included providing "urgently needed" housing at defense plants and military installations on a cost-effective basis, recovering the government's "entire investment" in Lustron, and "the establishment of a new and needed industry for the essential benefit of the entire nation."[88]

In Congress, the idea of reviving Lustron for defense housing production received considerable support. Representative Pat Sutton (R-TN), citing Lustron's excellent record in serving the military market, argued that "it is a proven fact that the type of house Lustron was building can be produced at a material saving over any other type of conventional house." Why throw away such an opportunity just because of some "shady dealings" between several people at Lustron and the RFC, Sutton asked. "Certainly now is no time to destroy a plant that is considered one of the finest of its kind in the country" for the "measly sum" of $5 million to be recovered through foreclosure.[89] Representative Earl Wilson (R-IN) echoed Sutton's argument and attacked the RFC and the Truman administration for its apparent unwillingness to follow through in its commitment to defense housing as outlined in Reorganization Plan No. 23. "Never was there a more crying need for housing facilities in view of our industrial and military expansion created by [the Korean War]."[90]

However, administration officials, prompted by the Joint Chiefs of Staff, favored transferring control of the Lustron plant to the Navy, which planned to lease it to the North American Aviation Co. for the production of carrier-based aircraft. Congressman Wilson considered the plan further evidence of "the Truman gang's waste and inefficiency" since North American was at least 1.5 years away from pilot production. "Mothers are writing to me about their sons . . . who are sleeping in tents in near zero weather," he said, while the administration refuses to support the production of defense housing. North American also did not plan to use the entire factory, which would leave approximately 900,000 square feet permanently idle. Why not at least share the space? "It is almost impossible to imagine the stupidity of a proposal to waste this floor space when it could be used to produce desperately needed defense housing," Wilson concluded.[91]

In February 1951 a subcommittee of the Senate Banking and Currency Committee entered the controversy over the disposition of the Lustron plant. Opening the hearings on February 13, subcommittee chairman John Sparkman (D-AL) reminded his colleagues that the nation was "confronted with the problems of getting housing during this mobilization period" and that he desired to "check out the situation at Lustron" as a potential solution. Sparkman was concerned that the transfer of the plant to the Navy was progressing with little consideration for its potential to provide defense housing. He did not deny the existence of sound reasons for the transfer but argued that "it ought not to be converted [to aircraft manufacture] unless the reasons were strong and overriding."[92]

Richard Dyas represented the RFC at the hearings and told the committee that while the agency was "energetically pursuing every lead," its hands were tied by the bankruptcy injunction issued by Judge Sullivan and affirmed by the Appeals Court. "If we had the property free and clear for disposal," Dyas said, "it could be leased on a basis which would provide a means of recovering the Government's investment." He also asserted his belief that the legal problems surrounding Lustron could be reconciled since most of the financial claims against Lustron's estate were based upon breech of contract arising out of the foreclosure action. If the RFC were free to act, whether in selling the plant to another concern or reorganizing Lustron, Dyas believed that the bankruptcy claims would be withdrawn on the basis of a continuation of operations. However, irrespective of the financial interests of the RFC, Dyas reminded the committee that the agency in January 1951 had referred the disposition matter to the Defense Production Administration (DPA). The DPA served as an arbiter among conflicting governmental interests in the Lustron plant. On January 30, the DPA ruled that the "defense effort would be better served if the plant were stripped of its machinery and the buildings turned over to the Navy for operation."[93]

The hearings on the disposition of Lustron's plant provided a forum for supporters of the company to voice their opposition to DPA's plans and an opportunity for Republicans to criticize Truman's policies on the issue of defense housing. They accomplished little more. Without evidence of overwhelming congressional opposition, Truman accepted DPA's suggestion prima facie. Indeed, despite a last-ditch effort on the part of Senator Sparkman to encourage the president to delay his decision, Truman on March 27, 1951, ordered the transfer of the Lustron plant to the Department of the Navy. "I am convinced that the decision was arrived at after careful study by all the agencies immediately concerned," Truman wrote in a telegram to Sparkman; "further delay would seriously retard overall aircraft production schedules established by the Joint Chiefs of Staff."[94] Any hope of salvaging Lustron as a going concern was lost.

In the aftermath of Truman's decision, the judiciary resolved Lustron's remaining legal encumbrances. On March 12 the Supreme Court declined to review the petition of appeal by the RFC and Lafayette Steel to overturn the decision of the Seventh Circuit Court of Appeals. The denial affirmed the Seventh Circuit opinion that Judge Sullivan's Chicago federal district court held peremptory jurisdiction over

Judge Underwood's Ohio federal district court. The Supreme Court's action thus terminated the foreclosure action and paved the way for the resolution of Lustron's remaining financial issues in federal bankruptcy court in Chicago.[95]

In July the court ordered the sale of Lustron's unmortgaged assets to satisfy the claims of the three Illinois dealers who had filed the original bankruptcy petition. A 129-page catalog listed over 4,000 lots of finished inventory, raw materials, and tools available at the Columbus plant. *Business Week* described the sale as a "windfall for some 400 bidders—purchasing agents, junk dealers, and housewives—who came from hundreds of miles around." The auction took in approximately $430,000, and the three dealers received payment for their claims of $4,842.[96]

In a broader context, by the summer of 1951 the furor over the Reconstruction Finance Corporation had also played out. The Fulbright committee fell victim to internal division and astute maneuvering by President Truman. Truman's Reorganization Plan No. 23 had addressed fundamental weaknesses in the RFC by implementing structural changes. A single administrator replaced the old board of directors, and the Plan established a Loan Policy Board to monitor the progress of firms receiving RFC funding. Additionally, in May 1951 Truman appointed Stuart Symington as RFC administrator, and Symington proceeded to "clean house." He implemented a public relations program designed to revive the reputation of the agency, including the public announcement each week of the amount and recipients of all RFC loans. Thus the Fulbright committee never formally recommended to the Senate that the RFC should or should not be abolished. The agency survived until July 30, 1953, when President Eisenhower signed the Small Business Act that officially terminated the RFC.[97]

In the interest of tying up the loose ends of the Lustron matter, in January 1952 the RFC dropped its suit against Carl Strandlund claiming liability on his personal guarantee of the original loan.[98] In 1953 Strandlund became president of Jerry O'Mahony, Inc., a manufacturer of "diner-type" restaurants. After several years there, he retired and moved to Florida. He spent the last years of his life in Minneapolis, Minnesota, and died there on December 24, 1974. His wife Clara told a *Minneapolis Tribune* reporter some years later that "it was his dream that every Joe could have a house . . . a lot of people have this misconception that he milked the government out of all this money, but no one could start a business of this kind and have them call in a loan in two years. He actually died of a broken heart."[99]

CHAPTER SIX

A Porcelain-Enameled Legacy

The best security for civilization is the dwelling, and upon proper and becoming dwellings depends more than anything else the improvement of mankind.
—Benjamin Disraeli

The Lustron Corporation retains a multifaceted legacy nearly six decades after Carl Strandlund first walked into Wilson Wyatt's office. Aided by the Internet and a heightened interest in the immediate postwar period, public awareness of Lustron has increased dramatically in recent years. Virtual communities of Lustron owners trade maintenance tips, post photographs, and share experiences of "the Lustron life" on numerous websites. Hundreds of articles in the local, regional, and national press have highlighted various aspects of the company's saga, and many television home improvement shows have mentioned Lustron in some manner. The company never lacked for publicity, of course, and Strandlund would no doubt appreciate the renewed enthusiasm and glowing comments about the physical legacy of his enterprise.

Solid As Steel

In 1994, Jim Morrow, a retired building-supply businessman and preservationist, opened his Lustron house in Chesterton, Indiana, near Gary, to the public as a "historic-house museum." It is currently one of several Lustron houses listed in the National Register of Historic Places. Along with the occasional help of Joe Tucker, Jr., who served as Lustron's director of quality control, Morrow conducts free public tours of the maize-yellow three-bedroom Westchester, number 2,329 off the Lustron production line. Publicity from several newspaper articles and the Indiana Bureau of Tourism enticed over 1,000 people to visit the house during the first three months of operation, and the house continues to attract a steady stream of tourists.[1]

Morrow purchased the house in 1991 for $52,000. The original owners, Norris and Harriet Coambs paid $13,516.55 for the house in 1949, which included a two-car Lustron garage. As Strandlund predicted, the house has held up remarkably well. Morrow repaired rusted roof tiles, a broken dishwasher/washing machine, and several damaged doors, but the porcelain-enameled panels remained in near pristine condition. "My Lustron house maintains a like-new condition," he reported. "Just the wipe of a damp cloth—no soap—removes anything." Period furnishings and decorations complement the interior. Morrow currently resides in the house, and despite some raised eyebrows in his family, maintains a high level of enthusiasm about his project. The "house is an important part of our heritage," he asserts, "and there's no sense keeping it a secret."[2]

The secret is indeed out. Jim Morrow's fascination with Lustron houses and the public's enthusiastic support of his endeavor reflects an ongoing interest in all things Lustron. Proud owners have populated the Internet with intricate websites detailing every aspect of their homes and corporate lore; Home and Garden Television has featured Lustron in its *Extreme Homes* series; an annual convention of Lustron homeowners began in 2000; and a major documentary film on Lustron has been shown on PBS affiliate stations throughout the nation. Even a causal web surfer would find over a thousand web pages containing references to Lustron in a simple query—not bad for a company that went out of business more than half a century ago.

Why the continuing interest? It's the house. There is still nothing quite like it, and anyone who has ever set foot in one is likely to remember the experience. From the beginning, the Lustron house received glowing reviews, and owners often became evangelists for the company.[3] Bernard E. Kinsock provided perhaps the best illustration from Lustron's operating era. As the company faced foreclosure in early 1950, Kinsock, an owner of a three-bedroom Westchester in Cuba City, Wisconsin, assailed the RFC and other opponents of the company for ignoring the high level of customer satisfaction frequently expressed among Lustron owners:

> I bought one of the first three-bedroom models in this section of the country . . . absolute satisfaction is my verdict on every count . . . unfavorable criticism is usually forthcoming from those who have never been inside a Lustron house.[4]

Kinsock's comments reflected a common concern among the original owners of Lustron houses. Many wondered why the government would withdraw its support of the company despite glowing reviews from its customers. "The Ford Motor Company did not get into full swing in one year," Kinsock noted. Delays in establishing marketing and service organizations seemed normal for any enterprise, especially one as ambitious as Lustron. Kinsock also wondered why the company's critics never mentioned "how the Lustron home provided a family with the best type of dwelling it could obtain [for the] price."[5]

Clearly the design and quality of Lustron's product demonstrated that many Americans would accept a factory-made house, a significant accomplishment given previous perceptions of prefabricated houses as cheap, impermanent "crisis" housing. In 1953, the U.S. Steel Corporation sponsored a survey of 320 Lustron owners to determine satisfaction with steel prefabricated housing and to provide research for a possible foray into the market. Two hundred owners returned the questionnaire, which covered such topics as age and income levels, length of occupancy, types of houses previously owned, and where they had first heard about the Lustron house. The survey also queried owners' opinions regarding design features such as the porcelain-enameled surfaces and mechanical features such as the heating system. It then asked them to compare the Lustron house with other prefabricated houses, other steel prefabricated houses, and, finally, conventionally built houses.[6]

Regarding the characteristics of Lustron owners, the survey found that 68 percent were between the ages of thirty and fifty and that 67 percent reported an annual income in excess of $5,000. Nearly 90 percent said that they were the original owner of the house, and 82 percent reported that they had lived in a conventionally built house before purchasing a Lustron. Validating the company's advertising strategy, 73 percent of the respondents said that they had first heard about the house from magazine advertisements or from press coverage of the company's model homes tour during 1948.[7]

The survey also revealed that a majority of owners, nearly 85 percent, believed that the house was large enough to suit their needs. Less than 10 percent had made structural changes, and the majority of those had enclosed the side porch. Intriguingly, only four of the 200 respondents reported altering the interior surfaces of their houses. Of the four, two had installed wood paneling in the living room and two had added wallpaper to the bathroom and kitchen.[8]

The 200 respondents reported high levels of satisfaction regarding the general livability and efficiency of their houses. Ninety-six percent said the house was warm in the winter, 90 percent said the house was cool in the summer, and 99 percent said the windows and doors were "weather tight." Moreover, nearly two-thirds of the owners reported a savings in heating costs over their previous, comparably sized houses. Over one-third said that they had saved as much as 30 percent in annual heating costs.[9] Yet the heating system also led the list of owner objections to the house. Fourteen percent reported difficulties in adjusting and maintaining uniform heating levels throughout the house. Other objections included the appearance of the asphalt tile baseboards (9 percent), the lack of materials to repair potential damage

A recently renovated Lustrom two-bedroom model.
Author's collection

(7.5 percent), improperly fitting doors (7 percent), and the lack of sufficient interior soundproofing (5.5 percent).[10]

In conclusion, the survey revealed that 97 percent of the respondents preferred the Lustron house to their previous houses for three main reasons: low maintenance costs, ease of cleaning, and convenient floor plan. Ninety-eight percent liked the overall design and the appearance of the porcelain-enameled surfaces, and 95 percent said they would purchase another Lustron house or a comparably featured steel house. When asked what changes they would suggest, 11 percent desired a larger kitchen, and 11 percent wanted access to the bathroom and bedrooms without going through the living room. Only 7.5 percent cited "expandability" as an issue that would influence the purchase of another steel prefabricated house.[11]

With such favorable reviews, the company would perhaps have had little trouble attracting customers, especially as it introduced new and more sophisticated models. In a continuing testament to the quality of the house, current Lustron owners report generally the same level of satisfaction at the beginning of the twenty-first century. Recently, Thomas Fetters and Vincent Kohler, founders of Lustron Research, a nonprofit organization dedicated to preserving the company's historical and physical legacy, estimated that "2,200 of the original 2,500 units originally produced are still standing and fully functional as low maintenance, pest-free homes." Lustron Research has documented approximately 2,000 houses and interviewed many of the current owners. Fetters and Kohler reported that only two current owners in their survey "expressed regret that they had acquired the home," which left "95.5 percent of the owners that were either inexpressive on their feelings or irrepressible in their enthusiasm for their Lustron."[12]

The existing houses have also attracted considerable attention from preservation societies as important artifacts of the postwar-built environment. In January 1996, the historic commission of the Chicago suburb Brookfield, Illinois, honored several local Lustron houses in a ceremony that united past and present owners. Mary Kircher of the commission commented that the houses "were unique to the American landscape" and a "legacy to the American people."[13] Another Lustron home in nearby Lemont, Illinois, caused a preservation controversy because it stood in the path of a proposed extension of Interstate 355, a tollway connecting Chicago's western suburbs. Homeowner Victor Pantaleo and the Will County historic preservation commission appealed to the Illinois State Tollway Authority to aid in securing funds to relocate the house. Pantaleo said, "Lustron houses really and truly are a unique part of our housing history. It's like rolling over Lincoln's home, but they don't care." The Tollway Commission eventually agreed to move Pantaleo's house to another site.[14]

Another threat to existing Lustrons is unfolding in established suburbs near major metropolitan areas where vibrant real estate markets have encouraged teardowns of modest postwar houses in favor of high-value new construction. The entry-level neighborhoods into which most original Lustron houses were placed are currently experiencing a different type of gentrification that threatens the viability of many houses of immediate postwar vintage. Some cities are taking steps to maintain the architectural integrity of original suburban developments by adopting new laws or highlighting the architectural history of the area. In Middleton, Wisconsin, the local landmarks commission recently undertook an educational project that included a Lustron house. The commission created and marketed jigsaw puzzles of historical local buildings to educate the public in the importance of the region's material heritage. One of the puzzles depicted a Lustron house located at 7120 North Avenue. Commission chairman Kent Calloway said the house was included to show people that "buildings don't have to be old to be unique."[15]

Lustron's legions of enthusiastic homeowners and others interested in the legacy of the enterprise have found a collective voice via the Internet and other mass media that might have helped the company survive its political travails. Lustron Research interviewed hundreds of current and former owners and reported widespread loyalty, devotion, enthusiasm, and pride in their houses. The "lust for Lustron" seems to transcend generations, as a number of young owners have created late-1940s time capsules in various rooms or in one case used the house as the centerpiece of a "kitschykastle" lifestyle. The line between kitsch and cool is indeed a fine one, and generation X seems to have embraced the house as an exemplar of both. Singer-songwriter Leonardo named one of his albums after the Lustron house, "not knowing the consequences of [his] actions." Soon he was receiving "emails requesting maintenance advice (I did help fix a roof once) . . . invitations to conventions, and interviews for documentary films . . . who knew the underbelly I had scratched."[16]

It is fair to speculate that had the Internet existed in 1950 it might have helped to rally supporters to the company's cause in a politically viable manner. However, even the Lustron dealers who expressed their desire to save the company before Congress ultimately had little influence over the firm's ultimate demise. Perhaps part of the company's continuing attraction is that its potential was left unfulfilled and its final chapter left unwritten, an asymmetric evolutionary arc that has enabled supporters and detractors alike free to speculate about what might have been.

What If . . .

Contemporary and subsequent evidence indicates that Lustron delivered a desirable, high-quality product to the housing market. Yet the company ultimately failed. Bernard Kinsock's assessment—that the single most important element of Lustron's failure was the government's impatience and shortsighted, politically motivated attitude toward the firm—spoke directly to the pitfalls of government entrepreneurship in the social realm. Since new enterprises of Lustron's scale are never the result of the efforts of a single individual, the quality, tenacity, and foresight of a firm's entrepreneurial network is crucial in transforming a promising idea into a commercial reality. Lustron's network included many energetic and tenacious constituents who believed in the firm and convinced others that something new and transformational was in the offing. Yet the federal government, Lustron's most important entrepreneurial partner, lost its taste for the venture and failed to foster additional networks and coalitions of resources to shepherd the enterprise toward maturity. Lustron supporter Representative Pat Sutton commented recently that Lustron might have survived the Washington jungle if it had employed well-connected lobbyists or reliable corporate representatives. Successful government entrepreneurs such as Henry Kaiser maintained effective Washington agents who arranged key meetings, smoothed ruffled feathers, and kept the company's finger on the governmental pulse.

The engineering consulting firm Booz, Allen & Hamilton, which authored a detailed report on Lustron's operations in 1949, concluded that the company had produced a high-quality house. However, the main issue for Lustron was not the technological challenges of design and manufacturing, but the process of motivating and binding heterogeneous constituencies to carry the infant enterprise toward stability. Once the crisis conditions that spawned Lustron had abated, the government abandoned the entrepreneurial partnership with devastating consequences. One of the most intriguing questions is whether a different political climate would have maintained the middle-tier strategy of federal support for innovative housing technologies, in the interest of improving the overall quality of housing for the mass market, while reducing costs to a point where even low-income Americans could afford decent single-family houses in suburban settings. Additional money and time might have produced a different reality for low- and moderate-income families.

Could Lustron have survived and prospered with additional time and money? The Booz, Allen report concluded that the average price of the house, around $10,000 including a lot, was simply too high for the mass market segment that Lustron had to serve in order to rationalize its tremendous productive capacity. The company recognized this problem and had intentionally geared its initial offering to a higher market segment because Strandlund and other top managers believed that only a full-featured house could overcome negative public perceptions of prefabricated housing. Lustron's early advertising consistently reminded customers that the Lustron house was not a "crisis" house, but a permanent, traditionally designed dwelling with more features and conveniences that a comparably priced conventionally built house. The company sought first to obtain public acceptance for its product, then expand the product line to serve broader market segments.

The initial prices for Lustron houses reflected greater than anticipated manufacturing expenses and costs of dealer assembly and site preparation that proved difficult to quantify given regional variations. Indeed Booz, Allen's analysis was correct—for 1949. The price of Lustron's early models limited the company's market potential and rendered inefficient its large investment in production capacity. But Lustron had by January 1950 undertaken a number of strategies to increase manufacturing efficiency and expand market potential.

The implementation of architect Carl Koch's plans for a "second generation" of Lustron houses represented the most important step. A reconfigured design emphasizing modularity promised to reduce raw material and manufacturing costs, speed the production process, and increase the number of models that the company could offer. For a relatively small investment in retooling and a revision of the production layout, Lustron was poised to offer models with an estimated cost ranging from $7,000 to $12,000, excluding lot. Profitable production of a $7,000 model represented the key to the wage earner market, while at the same time the company retained the capacity to serve higher market segments.

Koch's redesign would have rationalized Lustron's productive capacity and produced significant economic benefits for dealers and developers using Lustron houses. By increasing the level of factory preassembly, dealers and developers would spend less time and money to erect the house. Increased profitability promised a more stable distribution network and provided incentives to attract developers who could not otherwise afford to invest in training specialized crews to assemble Lustron houses. The "countrywide production line" envisioned by the company seemed within reach once profit potential could be assured throughout the production, distribution, and assembly process.

Further, by January 1950 Lustron had reevaluated its sales and marketing strategy and reconsidered its initial reliance on a franchised dealer network. The franchised dealer approach proved seductive as an initial strategy because it seemed to provide a flexible marketing structure. Variations in housing markets throughout the nation made it difficult for Lustron to devise strategies suited to each region. Local dealers seemed better equipped to implement tactical marketing strategies suited to local environments. Franchised dealers also relieved the company of the financial burden of carrying the cost of the house until it was sold to a customer.

Reliance on a franchise system, however, created a number of problems. The issue of control loomed large. In the automobile industry, for example, the push and pull of the dealer-company relationship had been resolved in favor of the companies, which exerted a high level of control over dealers with respect to pricing, inventory, and service requirements.[17] Lustron's system did not give the company sufficient control over its dealers to guarantee that a large production volume could be consistently moved onto the dealer network. Moreover, some dealers did not share the company's vision for the appropriate setting of the house. Individual dealers typically did not possess the financial resources to become community builders and thus focused on selling individual or small groups of houses that may or may not have been placed in community environments.

Consequently, Lustron implemented a general reorganization of its sales strategy that dropped all but the most profitable dealers, who—as it happened—had found success by implementing the community-planning suggestions contained in Lustron's Planning Guide. The company's emphasis on fleet sales—high-volume sales to suburban developers or to the military—promised to resolve the problem of unloading a large production volume while retaining control over the proper placement of the house within community environments. The "community building" marketing strategy included a commitment to expand factory support to large-scale developers and an extension of the financing arrangement with the Galbreath Mortgage Company to facilitate customer sales.

By establishing relationships with community builders instead of relying on small-scale franchised dealers as its market conduit, Lustron demonstrated an evolving understanding of its economic environment and an appreciation of the external factors that determined success or failure in the prefabricated housing industry. As William Levitt observed in 1949, "there is no such thing as a complete factory-engineered house—because no one has discovered how to prefabricate the land, how to prefabricate the road in front of the land, or the water main that goes into the house."[18]

Lustron indeed realized that only through a holistic approach to the process and consequences of prefabrication could it aspire to a profitable future. This was especially significant for a producer of factory-made housing. Firms like Levitt and Sons employed site prefabrication, a system in which the house was affixed to a community setting from the beginning of the construction process. For Lustron, extending administrative coordination beyond the shipping dock and encouraging partnerships with community builders represented a major step in the development of the industry. Whether the company's approach would ultimately have succeeded of course remains a matter of speculation.

As this study has argued, the firm's demise can be viewed as more a function of its relationship with the federal government than business shortcomings or market acceptance of its product. The company's reliance on government capital and association with the crisis atmosphere of the midcentury housing shortage placed it in a political crucible throughout its existence. Lustron's position reflected conflicts inherent in the process of government entrepreneurship in the social realm. Though the rewards may have been considerable, the drawbacks included incessant public attention and evaluation of even the smallest detail of corporate policy.

Lustron initially welcomed the attention as a valuable aspect of its advertising strategy because press coverage was overwhelmingly favorable. Yet such attention also created unrealistic expectations. When manufacturing and marketing difficulties delayed full-scale production, the media and government officials began to reconsider their support for Lustron. The terms of the relationship and the yardsticks for determining progress and success frequently shifted with the political winds. In five years Lustron's relationship with the government went from favorite son to unwanted stepchild.

The political environment that Lustron faced ultimately cost the company time, a commodity that in retrospect was as valuable as capital. Lustron's association with an ongoing debate regarding the proper role of the federal government in the postwar economy made it especially vulnerable to political attacks. The efficacy of federal capital allocation to industry in peacetime through the RFC became an issue that raised concerns about free competition and government domination of markets to the detriment of private enterprise. Many of Lustron's critics based their opposition solely on these ideological grounds. Other prefabricated housing firms also raised the issue of anticompetition, noting that government "favoritism" of Lustron created a barrier to entry for firms that were not fortunate enough to receive federal capital.

The tenuous political position of the RFC also undermined Lustron's prospects. As the agency battled for its continued existence, its association with Lustron became a political liability. The terms of the Lustron loans reflect the RFC's desire to plot a course between supporting a promising venture that served public policy goals, and the temptation to abandon Lustron in order to shore up its political position. Recognizing the complexity of the venture, the original loans granted a long-term repayment schedule. But the later loans required full repayment within a time span as short as six months. The short repayment terms of the later loans thus provided the RFC with a rationale for terminating its association with Lustron.

Without access to long-term capital, the company's plans to refine production and marketing strategies could not be realized. Ironically, one of the most frequently cited reasons for optimism regarding Lustron's success was its large capitalization that enabled the attainment of production economies. In the end, however, Lustron was highly undercapitalized as a result of the withdrawal of RFC support. The company was thus unable to exploit the competitive advantage of scale, a factor that had long been viewed as essential to the rationalization of the American housing industry.

Clearly, both the RFC and Lustron made many mistakes. Lustron's management team grew accustomed to special treatment and at times adopted an arrogance that exacerbated difficulties with the government. Strandlund admitted that Lustron was "in the habit of getting what it want[ed]." This attitude fostered resentment among its critics and sometimes even its supporters as the company expected them to run interference when bureaucratic roadblocks appeared. The RFC was also inconsistent in defining the limitations of its involvement in company affairs. Initially adopting a hands-off attitude, as political pressures increased and business problems inevitably emerged, the RFC became interested in minute operational details. The increased level of interest was undoubtedly a political defense mechanism as it sought to justify its considerable investment. Yet to Lustron officials it appeared that the RFC was simply looking for shortcomings and problems to use as ammunition in a quest to destroy the company. The lack of communication between the company and its primary creditor produced an atmosphere of mistrust that obviated what otherwise might have been a beneficial aspect of the company's relationship with the federal government.

Contemporary observers of the immediate postwar housing industry often called for "revolutionary" approaches to facilitate the mass production of affordable housing. Lustron was indeed revolutionary, particularly with respect to the scale and scope of its operations and the level of federal support that it initially received. It may, however, have been too radical. Not from a product perspective, for the market acceptance of the Lustron house demonstrated that the public would buy a porcelain-enameled steel dwelling, but from a process perspective.

The unprecedented level of direct federal financial participation in one company represented a significant departure from traditional peacetime housing policies. Despite the crisis atmosphere created by the veterans' housing issue, such an investment seemed out of place within a peacetime economic context. Conservatives especially viewed federal support for Lustron as inconsistent with their desires for reduced governmental involvement in the private sector. Indeed, the Housing Act of 1949 included the objective that "private enterprise shall be encouraged to serve as large a part of the total need as it can" and "governmental assistance shall be utilized where feasible to enable private enterprise to serve more of the total need."[19] Opponents of Lustron clearly believed that continued governmental assistance was not a "feasible" element of federal housing policy, and the American public ultimately accepted this analysis.

Legacies

Though not as evident as the houses themselves, other elements of Lustron's legacy remain relevant today. The technological scope of the firm was significant and influential. The allure of factory-built housing did not disappear after Lustron, and in subsequent years the housing industry has embraced prefabricated technologies with substantial vigor. Instead of focusing on one firm or one type of technology, though, the industry has made steady but sure progress across a range of processes and materials. This has allowed it to digest the impact of industrialization on market practices and business systems in a more comfortable and contextual manner. At the dawn of a new century, however, prefabricated housing technologies can be found in virtually all newly constructed single-family houses. Even in high-value "custom-built" houses elements of industrialized building are used with increasing frequency. Calling the average million-dollar home "prefabricated" is of course a stretch, but in my suburban Chicago neighborhood "conventionally" built houses using factory-made modules can sell just a few price points south of the million-dollar mark.

Industrialization is now a reality in residential construction across a wide range of housing forms. Perhaps the most familiar is the mobile home, also called the manufactured house, since it is rarely mobile in its present use. Despite a legacy of criticism and condescension, manufactured housing plays an important role in the housing market as the "predominant unsubsidized type of affordable housing in the United States." The manufactured housing industry is especially important in providing decent housing in less populated areas, but its impact is felt nationwide across a surprising demographic range. Over 12.5 million people live in manufactured houses, and the industry is growing at nearly twice the rate as the conventional homebuilding industry.[20]

Part of the success of manufactured housing has been the consistent development of larger models with more sophisticated features and higher-quality materials engineered to surprisingly high tolerances. Another factor has been its increasing use outside the "trailer park" in traditional, desirable neighborhood settings. There is a good chance that many Americans have driven by a "mobile home" without realizing it. Manufactured housing, with a little help from a landscape designer and finish carpenter, has proven well suited to many neighborhoods and continues to offer value to a market segment frequently neglected by today's merchant builders. Indeed, the manufactured housing industry has increasingly demonstrated an understanding of the total dimensions of industrialized housing and has provided valuable lessons to the industry as a whole.

The increasing sophistication of the manufactured housing industry in recent decades has indeed provided greater opportunity for home ownership in the lower-income market. Yet the Lustron house of the late 1940s remains superior to most manufactured houses currently on the market. Lustron's engineering and build quality far exceeded most contemporary manufactured homes in comparable market segments. In contrast to the surveys of Lustron owners cited above, owners of manufactured housing report wide variations in quality and satisfaction. The extensive use of environmentally unsustainable products in manufactured housing and their associated deleterious health effects is but one of the social costs of abandoning Lustron's more sophisticated, integrated approach. Given the firm's engineering expertise and dedication to quality and innovation, there is little doubt that a twenty-first-century Lustron house would be superior to any contemporary manufactured home and would exert considerable influence over build quality and cost structure across the market spectrum of modern housing construction.

Serving traditionally underserved markets through the economic benefits of industrialized production has been a focus of those concerned with American housing issues since the late-nineteenth century. Yet the impetus for industrialization today reaches well beyond concern for the underserved to encompass all sectors of the residential construction industry. As demand pressures force housing costs higher, the need for greater efficiency and productivity is clear. Industry forecasters predict a housing boom while at the same time warning of a continuing shortage of carpenters and trades people over the next decade. Technology, in the form of improved production processes or new materials, has gained increasing attention across the industry in recent years.

The rise of environmentalism has also emerged as a major issue for the housing industry and a motivation for increasing the efficiency of the production stream. Prefabricated components and systems help reduce waste at the building site and thus help firms reduce significant waste management costs. Rising prices and declining quality of lumber have prompted the industry to give building materials other than wood serious consideration. A number of companies in the steel and construction industries have promoted the use of steel in residential construction as a response to the increasing environmental consciousness among consumers. Several companies cited the innovative quality of Lustron's steel frame design at a recent convention of the National Association of Home Builders in Houston, Texas. Steel framing systems consisting of roof trusses and wall studs displayed at the convention bore a remarkable resemblance to the Lustron system, and firms touted the same advantages of steel framing, such as durability and structural stability, as Lustron did in 1947.[21]

The Steel Framing Alliance, an organization established by the American Iron and Steel Institute in 1998, continues vigorously to promote steel-framed housing. The steel industry's longstanding interest in housing as an additional market for its products is well served by contemporary environmental concerns. As one member observed, "steel-frame housing gives steelmakers a chance to portray themselves as environmentalists. After all, the choice between wood and steel for the frame of an average house is the choice between one acre of trees . . . or seven Oldsmobiles." The Alliance believes that steel is the best and most feasible alternative building material for residential construction and that it has "inherent benefits that can be found in no other product."[22]

Additional advantages of steel construction touted by the Alliance would sound familiar to Lustron's sales staff: impervious to termite damage; noncombustible; dimensionally stable; consistent material quality; screwed and bolted together for greater resistance to wind damage; not vulnerable to any type of fungi or organism; and, interestingly, "reduces work for other trades as it has pre-punched holes for running piping and electrical wiring."[23] Steel framing and steel housing systems are also attractive from a long-term performance and quality standpoint, especially since insurance companies are increasingly wary of litigation surrounding building component warranties on conventionally build housing developments. Clearly, wood is no longer automatically seen as the best material for housing, and this has forced builders to become more familiar with alternative systems and aware of possible competitive inroads.

Other factors currently influencing the advance of industrialization is the ongoing consolidation of the building industry. Big builders are getting bigger, and the increased geographical scope of individual firms has heightened competition. The scale and scope of these firms and competitive pressures associated with the pursuit of new markets have made them more aware of the advantages of factory-built components, particularly in controlling the cost of site labor. Interestingly, as these firms become more powerful they have been able to influence the adoption of performance-based building codes and have thus paved the way for further prefabrication in wood and other materials amenable to factory production.

While current market forces seem to favor the advance of housing technologies, one factor that has had minimal influence on industrialization in the decades following Lustron's demise is federal housing policy. This is primarily a consequence of the triumph of privatism and prosperity in the postwar era. The "path not taken"—providing direct support for innovative housing firms—meant that the two-tiered housing policy that emerged in the 1930s remained preeminent. The advance of housing technologies was ancillary to concerns over the continuation of subsidies for the private construction of suburban-based middle-class housing on one hand and publicly funded urban housing projects for low-income families on the other. Despite the importance of the housing industry to the American economy, there has been very little government investment or interest in residential technologies. The federal government bears substantial blame for the continuing fragmentation of the industry and for inhibiting the introduction of new technologies. Many emerging technologies rarely get past the prototype stage since the federal government remains unwilling to offer sufficient incentives to support experiments beyond initial stages of research, development, and market testing. Strandlund would have understood.

However, beginning in the 1990s, the use of industrialized housing to achieve public policy goals has again attracted the attention of government policy makers. In 1990 Congress passed the Cranston-Gonzalez National Affordable Housing Act that established the National Commission on Manufactured Housing. The Commission called for "equal treatment" of prefabricated houses with respect to financing and federal housing subsidies.[24] The department of Housing and Urban Development has also supported the use of manufactured housing to provide "elder cottages" for low-income seniors. One hundred units in seven cities have been erected under the Elder Cottage Housing Opportunity program.[25] In 1994 Congress passed the Veterans' Benefits Improvements Act, which included Veterans Administration loan guarantees for manufactured housing.[26] In this respect, the Act is reminiscent of the Veterans' Emergency Housing Act of 1946, the legislation that aided the establishment of the Lustron Corporation.

Industrialized housing also continues to attract attention as a strategy to provide affordable housing and economic opportunity. A current focus is on economically depressed regions in major cities. The Automated Builders Consortium, founded in Los Angeles by Don Carlson in 1992, envisioned building prefabricated housing plants in the inner city. Carlson stated that

> it occurred to us that we ought to start building housing factories in the inner city that would employ local people and teach them skills . . . the more you can do in

> a factory, the more you have an opportunity to train people and the less costly the product will be.

Echoing the theme of government entrepreneurship, Carlson planned to "take advantage of government financing programs . . . and involve the banks and mortgage companies." Several factories are currently in operation, including two in Los Angeles, one in Wichita, Kansas, and one on the south side of Chicago.[27]

These firms will undoubtedly face some of the same problems that Lustron did half a century earlier. Like Lustron, it appears that the current ventures have a strong commitment to providing a quality product. As Jim Kuhnert of New Era Building Systems recently observed, "we build to the top of the code . . . I think our product is better than the average site-built house."[28] Perhaps their customers will express the same level of satisfaction with their houses fifty years from now, as many Lustron owners have since 1949.

Another significant development in the area of government–industry cooperation is the creation of PATH, the Partnership for Advancing Technology in Housing. A national effort to improve the housing design and construction industry, PATH began in 1994 when the Clinton administration created a clearinghouse organization to assist a variety of federal agencies and private participants in the formulation of national construction goals. The effort heightened interest in an ongoing public–private partnership, and in 1998 Congress officially authorized the organization to foster the development, dissemination, and use of new housing technologies under the coordination of the Department of Housing and Urban Development. PATH addresses "issues and institutional problems related to technology development in the housing industry . . . strives for viable cost-effective solutions" and has "adopted far-reaching goals that deal with the quality, durability, environmental impact, energy efficiency, affordability, and disaster risk of America's homes." With partners from all sectors of the residential construction industry, PATH seeks to meet three primary objectives: developing new housing technologies; disseminating information about new and existing housing technologies; and studying and establishing mechanisms for sustained housing technology development and market acceptance.[29]

PATH clearly holds promise as a model for business and industry collaboration outside traditional boundaries. Expenditures on residential technology research and development continue to languish as a percentage of sales in comparison to other industries. Emerging technologies that could be useful to the housing industry often fail to find acceptance because the housing industry, unlike other driving forces in our economy, is extremely dispersed. The introduction of new technologies can take decades to achieve full market penetration. By helping industry to close these gaps and to get next-generation technology into the market faster, PATH hopes to have a significant impact on improving the quality and affordability of American housing.[30]

The recently issued Millennial Housing Commission report "Meeting Our Nation's Housing Challenges" contains other elements suggesting renewed federal interest in housing technologies and prefabricated construction. The report affirmed the importance of manufactured housing to the American housing system and called for the expansion of credit access for buyers of manufactured houses. The report also asked Congress to affirm that Fannie Mae and Freddie Mac can purchase loans on manufactured homes classified as personal property, encouraged support of a secondary market in such loans if they are determined to be sound, and established performance goals for manufactured home loan purchases. Integrating all types of homes into the U.S. housing finance system on an equitable basis is an important element for the future of industrialized housing. The report also generally affirmed the importance of government-sponsored enterprises such as Fannie Mae, Freddie Mac, and the Federal Home Loan Banks in the development of industrialized housing to serve low- and moderate-income homebuyers.[31]

Whether these programs will hasten the process of industrialization remains to be seen. With the aid of market forces favorable to prefabrication, they may produce technological advances that would aid all sectors of the industry. It is unlikely, however, that another venture like the Lustron Corporation will emerge from the cauldron of government–industry interaction. Lustron was the product of a line of evolution within the housing industry unceremoniously extinguished by political and economic circumstances. And that was a loss for society, since with few exceptions the provision of housing for ordinary Americans has received scant attention in subsequent decades. The concerted energies of the Truman administration, Congressmen from both parties, and various administrators in dealing with housing in the immediate postwar years must be recognized as one of the few eras when actions as well as words were aimed at easing the nation's enduring housing problems. Wyatt, Truman, and other supporters of Lustron's vision for a new era in housing production deserve credit for risking their reputations in the quest to find solutions to assist Americans of modest means. They challenged long-established and powerful forces in the conventional home building industry and demonstrated an awareness of historical influences and future possibilities. In the end they lost the battle, but they remain largely unsung heroes in the ongoing campaign for the fulfillment of the national housing goals as expressed in the Housing Act of 1949 of "a decent home and living environment for every American family."

Unlike the majority of failed business enterprises, Lustron left considerable tangible evidence of its existence. Visitors to Jim Morrow's Lustron "museum" and current owners of Lustron houses can still share a part of Carl Strandlund's vision. It is likely that the Lustron house will endure as a small but intriguing part of America's material heritage for some time to come. In 1959, C. W. Webster of 2021 Kilburn Avenue in Rockford, Illinois, told a reporter about the time a runaway truck smashed into the corner of his Lustron house. "The truck was completely demolished," he said, "but two former Lustron carpenters repaired the damage to the house in a day and a half at a cost of approximately $150." The carpenters stated that "any other house would have been knocked off its foundation by the force of the collision."[32] The Lustron house at 2021 Kilburn Avenue remains standing today.

NOTES

Notes to Introduction

1. "The Factory Built House Is Here, But Not the Answer to the $33 Million Question: How to Get It to Market?" *Architectural Forum* 90 (May 1949): 107.

2. John Morton Blum, *V Was For Victory: Politics and American Culture during World War II* (New York: Harcourt, Brace Jovanovich, 1976), 102.

3. The government defined a prefabricated house in 1947 as "one having walls, partitions, floors, ceilings, and/or roof composed of sections or panels . . . which have been fabricated in a factory prior to erection on the building foundation. This is in contrast to the conventionally built home which is constructed piece by piece on the site." Quoted in Prefabricated Home Manufacturers' Institute and U.S. Department of Commerce, *Prefabricated Homes, Commercial Standard CS125–47* (Washington, D.C.: U.S. Government Printing Office, 1947), 1.

4. Burnham Kelly, *The Prefabrication of Houses: A Study by the Albert Farwell Bemis Foundation of the Prefabrication Industry in the United States* (Cambridge, MA: Technology Press of the Massachusetts Institute of Technology, 1951), chapter 1. Alfred Bruce and Harold Sandbank, *A History of Prefabrication* (New York: Arno Press, 1972), 6–7. Joseph B. Mason, *History of Housing in the U.S., 1930–1980* (Houston: Gulf Publishing Company, 1982), 54–55. Conceptualizing the production of housing as a single integrated process opened the door to numerous potential advantages. Standardization, repetition, specialization, and efficiencies in material handling represented the primary benefits of mass production. The prospect of producing a higher-quality product with less-skilled labor bolstered the argument that prefabrication would offer consumers a better house for less money than conventional construction. Conventional building required a labor force that was approximately 60 percent skilled, 5 percent semiskilled, and 35 percent unskilled. In contrast, factory-made housing used approximately 25 percent skilled, 35 percent semiskilled, and 40 percent unskilled labor. Advantages of scale, particularly the ability to obtain volume discounts from suppliers, held additional economic promise. Further, the idea of freeing the home-building process from its seasonal cycle, with the promise of overall higher wages for labor, confirmed the appeal of relocating housing production in the factory.

5. Quoted in Richard Bender, *A Crack in the Rear-View Mirror: A View of Industrialized Building* (New York: Van Nostrand Reinhold, 1973), 6.

6. C. Theodore Larson, "Toward a Science of Housing," *Scientific Monthly* 65 (October 1947): 295–305. Charles Lewis Owen, *Design and the Industrialized House* (Chicago: Institute of Design Press, Illinois Institute of Technology, 1965), 19–49. Gilbert Herbert, *The Dream of the Factory-Made House: Walter Gropius and Konrad Wachsmann* (Cambridge, MA: MIT Press, 1984), 7. Herbert points out that architects were beginning to understand the concept of "system" during the 1920s and 1930s due to the influence of systematic thought in other disciplines: "Jan Christian Smuts published his philosophy of holism, and Alfred North Whitehead his analogous theory of organic mechanism. Ludwig

von Bertalanffy laid the foundations of a general systems theory, and Arthur Tansley applied parallel ideas to the environment in his concept of the ecosystem."

7. Quoted in Bemis, *The Evolving House: The Economics of Shelter (Cambridge,* 1934), 335.

8. From the turn of the century through World War II, America's housing output had not kept up with the expanding population, and the nation's housing had progressively deteriorated. According to the 1940 census, the 34,834,532 dwellings in the United States had a median age of over 25 years, and large percentages lacked running water, private baths, flush toilets, and refrigeration. The largest annual production occurred in 1925 (937,000 units) and the lowest occurred in 1933 (93,000 units). Many critics cited technological backwardness as the primary reason for the inability of the housing industry to keep up with demand.

9. Stephen B. Adams, *Mr. Kaiser Goes to Washington: The Rise of a Government Entrepreneur* (Chapel Hill: University of North Carolina Press, 1997), 1–13. See also Adams, "Mr. Kaiser Goes to Washington: Saga of a Government Entrepreneur" (Ph.D. diss., Johns Hopkins University, 1994), introduction.

10. United States Bureau of the Census, *Statistical Abstract of the United States: 1948* (Washington, D.C.: U.S. Government Printing Office, 1948), 783.

11. Bemis, *The Evolving House,* 144.

12. "That Lustron Affair," *Fortune* 40 (November 1949): 94.

13. Kenneth Lipartito has pointed out the dangers of viewing the business process apart from its sociocultural setting. Kenneth Lipartito, "Culture and the Practice of Business History," *Business and Economic History* 24 (Winter 1995): 1–41.

14. Robert B. Fairbanks, *Making Better Citizens: Housing Reform and the Community Development Strategy in Cincinnati, 1890–1960* (Urbana: University of Illinois Press, 1988), 1–3. Kristin Bailey, "The Federal Government and the Cooperative Housing Movement, 1917–1955" (Ph.D. diss., Carnegie Mellon University, 1988), introduction. *United States Statutes at Large* 63, pt. 1, 413.

15. Douglas Haskell, "Prefabrication: Assembly Lines Reach Out for New Markets," *Architectural Record* 93 (June 1943): 63. Albert C. Schweizer, "Factory-Built Houses an Urban Threat," *The American City* 58 (December 1943): 64.

16. "The Lustron Planning Guide," Lustron Corporation Records, Ohio Historical Society, Columbus, Ohio, Box 2, Folder 3. Hereinafter cited as LCR.

17. Marc A. Weiss, *The Rise of the Community Builders: The American Real Estate Industry and Urban Land Planning* (New York: Columbia University Press, 1987), 2–3.

18. Ibid., 3.

19. Kelly, *The Prefabrication of Houses,* 289–96, 336.

20. William H. Whyte's bestseller *The Organization Man* (New York: Simon and Schuster, 1956) used Park Forest to illustrate patterns of suburban development and their perceived effects on human social and political behavior.

21. Examples of wartime prefabrication that reinforced negative perceptions included mobile homes, trailers, and Quonset huts. Although these dwellings performed admirably as temporary defense housing, few people had a Quonset hut in mind as the ideal of postwar homeownership. Zoning restrictions on "demountable" (temporary and transportable) prefabricated houses appeared during World War II in response to public distaste for "minimum standard" prefabrication erected in high density patterns near defense and military installations. A consumer opinion poll in August 1944 revealed that while 74.5 percent of those interviewed had heard of prefabricated houses, only 17.2 percent

would consider living in one permanently. The most frequent reasons given for not wanting a prefabricated dwelling were "lack of strength" and "short life." Kelly, *The Prefabrication of Houses,* 62–63. "The *Fortune* Survey," *Fortune* 33 (April 1946): 275.

22. Thomas S. Dicke, *Franchising in America: The Development of a Business Method, 1840–1980* (Chapel Hill: University of North Carolina Press, 1992), argued that control over the behavior of franchisees was a crucial element in the success of the product-franchising marketing strategy. Lustron struggled to gain control of its franchise dealer network and bemoaned the tendency of their dealers to desire large profits on fewer sales rather than smaller profits on greater sales volume. The lack of control over dealers resulted in higher prices to consumers and reduced the company's ability to serve a mass market.

23. Sherman J. Maisel, "Policy Problems in Expanding the Private Housing Market," *The American Economic Review* 41 (May 1951): 599–611.

24. Alfred P. Sloan, *Adventures of a White Collar Man* (New York: Doubleday, Doran & Company, 1941), 199.

25. "Porcelain Enamel Homes Have 27-Year History," LCR, Box 1, Folder 7.

Notes to Chapter 1

1. Kenneth T. Jackson, *Crabgrass Frontier: The Suburbanization of the United States* (New York: Oxford University Press, 1985). Gwendolyn Wright, *Building the Dream: A Social History of Housing in America* (New York: Pantheon Books, 1981). Stuart Blumin, *The Emergence of the Middle Class: Social Experience in the American City, 1760–1900* (Cambridge: Cambridge University Press, 1989), 138–40.

2. Roy Lubove, *The Progressives and the Slums: Tenement House Reform in New York City, 1890–1917* (Pittsburgh: University of Pittsburgh Press, 1962). Alan F. Davis, *Spearheads for Reform: The Social Settlement and the Progressive Movement, 1890–1914* (New York, 1967). Robert H. Wiebe, *The Search for Order, 1877–1920* (New York: Hill and Wang, 1967).

3. Lubove, *The Progressives and the Slums.* Robert H. Bremner, *From the Depths: The Discovery of Poverty in the United States* (New York: New York University Press, 1956). Eugenie Birch and Deborah Gardner, "The Five Percent Solution: A Review of Philanthropic Housing, 1870–1910," *Journal of Urban History* (August 1981): 408–38.

4. The first federal action specifically addressing housing was a Congressional resolution in 1892 that provided $20,000 for investigation of slum conditions in cities with populations exceeding 200,000. Paul Boyer, *Urban Masses and Moral Order in America, 1820–1920* (Cambridge, MA: Harvard University Press, 1978). Richard Plunz, *A History of Housing in New York City: Dwelling Type and Social Change in the American Metropolis* (New York: Columbia University Press, 1990). Lubove, *The Progressives and the Slums.*

5. Margaret Crawford, *Building the Workingman's Paradise: The Design of American Company Towns* (New York: Verso, 1995).

6. Ibid.

7. Robert B. Fairbanks, "From Better Dwellings to Better Neighborhoods: The Rise and Fall of the First National Housing Movement," in *From Tenements to the Taylor Homes: In Search of an Urban Housing Policy in Twentieth-Century America, ed.* John F. Bauman, Roger Biles, and Kristin Szylvian (University Park: Pennsylvania State University Press, 2000), 21–42. Fairbanks, *Making Better Citizens: Housing Reform and the Community Development Strategy in Cincinnati, 1890–1960* (Urbana, 1988). Roy Lubove, ed., *The Urban Community: Housing and Planning in the Progressive Era* (Englewood Cliffs, NJ: Prentice-Hall, 1967).

Robert G. Barrows, "Beyond the Tenement: Patterns of American Urban Housing, 1870–1930," *Journal of Urban History* 9 (August 1983): 395–420. Howard Gillette, "The Evolution of Neighborhood Planning from the Progressive Era to the Housing Act of 1949," *Journal of Urban History* 9, 4 (1983): 421–44.

8. Crawford, *Building the Workingman's Paradise.* Stanley Buder, *Pullman: An Experiment in Industrial Order and Community Planning, 1890–1930* (New York: Oxford University Press, 1967). Anne Mosher, "Capital Transformation and the Restructuring of Space" (Ph.D. diss., The Pennsylvania State University, 1989).

9. Eric J. Karolak, "'No Idea of Doing Anything Wonderful': The Labor-Crisis Origins of National Housing Policy and the Reconstruction of the Working-Class Community, 1917–1919," in *From Tenements to the Taylor Homes: In Search of an Urban Housing Policy in Twentieth-Century America,* ed. John F. Bauman, Roger Biles, and Kristin Szylvian (University Park: Pennsylvania State University Press, 2000), 60–80. Kristin Szylvian, "The Industrial Housing Reform and the Emergency Fleet Corporation," *Journal of Urban History* 125, 5 (1999): 647–90. Christian Topalov, "Scientific Planning and the Ordering of Daily Life: The First 'War Housing' Experiment in the United States, 1917–1919," *Journal of Urban History* 17, 1 (1990): 14–45.

10. Ibid. The Emergency Fleet Corporation operated by providing loans to real estate companies incorporated by shipbuilding companies. The U.S. Housing Corporation directly organized, built, and managed its communities. Curtice N. Hitchcock, "The War Housing Program and Its Future," *Journal of Political Economy* 27 (April 1919): 241–79.

11. Karolak, "'No Idea of Doing Anything Wonderful,'" 70–75.

12. Ibid., 70–75. John S. Garner, "The Garden City and Planned Industrial Suburbs: Housing and Planning on the Eve of World War I," in *From Tenements to the Taylor Homes: In Search of an Urban Housing Policy in Twentieth-Century America, ed.* John F. Bauman, Roger Biles, and Kristin Szylvian (University Park: Pennsylvania State University Press, 2000), 43–59. Richard Candee, *Atlantic Heights: A World War I Shipbuilder's Community* (Portsmouth, NH: P. E. Randall, 1985).

13. Karolak, 74–77.

14. Janet Hutchinson, "Shaping Housing and Enhancing Consumption: Hoover's Interwar Housing Policy," in *From Tenements to the Taylor Homes: In Search of an Urban Housing Policy in Twentieth-Century America, ed.* John F. Bauman, Roger Biles, and Kristin Szylvian (University Park: Pennsylvania State University Press, 2000), 81–101. Ellis Hawley, "Herbert Hoover, the Commerce Secretariat, and the Vision of the Associative State," *Journal of American History* 61 (June 1974): 116–40.

15. Hutchinson, "Shaping Housing and Enhancing Consumption," 85–95. Hutchinson, "Building for Babbitt: The State and the Suburban Home Ideal," *Journal of Policy History* 9 (1997): 184–210. Carolyn S. Loeb, *Entrepreneurial Vernacular: Developers' Subdivisions in the 1920s* (Baltimore: Johns Hopkins University Press, 2001), chapter 4.

16. Ibid.

17. Ibid.

18. Ibid., 95–96. Gilbert Herbert, *The Dream of the Factory-Made House: Walter Gropius and Konrad Wachsmann* (Cambridge, MA: MIT Press, 1984), introduction.

19. Hutchinson, "Shaping Housing and Enhancing Consumption," 96. Hutchinson, "Building for Babbitt," 208–10. Michael J. Doucet and John C. Weaver, "Material Culture and the North American House: The Era of the Common Man, 1870–1920," *The Journal of American History* 72 (December 1985): 560–87.

20. The quotation is from an influential article in *Fortune* magazine in 1947 chroni-

cling the inefficiency of the housing industry. *Fortune* was consistently harsh on the industry and gave industry pioneers and visionaries considerable positive press. The subtitle of the article sums up the editorial board's frustrations: "Only major money and modern organization, plus brains, will ever rescue the housebuilding business from its feudal controls and its chronic incompetence." "The Industry Capitalism Forgot," *Fortune* 36 (August 1947): 61–67.

21. Edward Filene, *The Way Out: A Forecast of Coming Change in Business and Industry* (Garden City, NY: Page & Company, 1925).

22. Lewis Mumford, *City Development: Studies in Disintegration and Renewal* (New York: Harcourt, Brace and Company, 1945). Robert Fishman, *Bourgeois Utopias: The Rise and Fall of Suburbia* (New York: Basic Books, 1987).

23. Ibid. Roy Lubove, *Community Planning in the 1920s: The Contribution of the Regional Planning Association of America* (Pittsburgh, 1963). Mark Baldassare, "Suburban Communities," *Annual Review of Sociology* 18 (1992): 475–94.

24. Rosalyn Baxandall and Elizabeth Ewen, *Picture Windows: How the Suburbs Happened* (New York, 2000), 46–48. Daniel Schaffer, *Garden Cities for America: The Radburn Experience* (Philadelphia: Temple University Press, 1982).

25. Mumford, "America Can't Have Housing," *The New Republic* (1934): 15–19.

26. Catherine Bauer, *Modern Housing* (New York: Houghton Mifflin Company, 1934), 242.

27. Burnham Kelly, *The Prefabrication of Houses: A Study by the Albert Farwell Bemis Foundation of the Prefabrication Industry in the United States* (Cambridge, MA: Technology Press of the Massachusetts Institute of Technology, 1951), 12–14.

28. Ibid., 15–20. U.S. National Resources Committee, *Technological Trends and National Policy, Including the Social Implications of New Inventions* (Washington, D.C.: U.S. Government Printing Office, 1937), 370–75.

29. Kelly, *The Prefabrication of Houses,* 21–24. John S. Burchard, "Research Findings of Bemis Industries, Inc.," *Architectural Record* 75 (January 1934): 3–8.

30. Ibid.

31. Ibid.

32. Kelly, *The Prefabrication of Houses,* 26–28. "Housing VI: Solutions," *Fortune* 6 (July 1932): 60–68. Buckminster Fuller, "Dymaxion House: An Attitude," *Architectural Record* 75 (January 1934): 10. Fuller, *Designing a New Industry: A Composite of a Series of Talks, 1945–1946* (Wichita: Fuller Research Institute, 1946). "Fuller's House," *Fortune* 33 (April 1946): 167.

33. Quoted in Baxandall and Ewen, *Picture Windows,* 21–22.

34. After 1925, the residential construction industry had suffered a steady decline, from $4.5 billion in 1925 to $2.4 billion in 1929. Nathaniel Keith, *Politics and the Housing Crisis since 1930* (New York: Universe Books, 1973), 17–19. Richard O. Davies, *Housing Reform during the Truman Administration* (Columbia: University of Missouri Press, 1966).

35. Baxandall and Ewen, *Picture Windows,* 51–54. Gail Radford, "The Federal Government and Housing during the Great Depression," in *From Tenements to the Taylor Homes: In Search of an Urban Housing Policy in Twentieth-Century America, ed.* John F. Bauman, Roger Biles, and Kristin Szylvian (University Park: Pennsylvania State University Press, 2000), 102–20. Gail Radford, *Modern Housing for America: Policy Struggles in the New Deal Era* (Chicago: University of Chicago Press, 1996).

36. Ibid.

37. Ibid.

38. Marc A. Weiss, *The Rise of the Community Builders: The American Real Estate Industry*

and Urban Land Planning (New York: Columbia University Press, 1987), chapter 6. Radford, "The Federal Government and Housing," 115. Arthur M. Weimer, "The Work of the Federal Housing Administration," *Journal of Political Economy* 45 (August 1937): 466–83.

39. Baxandall and Ewen, *Picture Windows,* 54–56.

40. Baxandall and Ewen, 59–60. Kelly, *The Prefabrication of Houses,* 36–37. Charles Abrams, "The Ft. Wayne Plan Analyzed," *American City* 54 (April 1939): 106. Eric J. Sandeen, "The Design of Public Housing in the New Deal: Oskar Stonorov and the Carl Mackley Houses," *American Quarterly* 37 (Winter 1985): 645–67.

41. Baxandall and Ewen, 60–61. Radford, "The Federal Government and Housing," 109–10.

42. Baxandall and Ewen, 62–66. Timothy McDonnell, *The Wagner Housing Act: A Case Study of the Legislative Process* (Chicago: Loyola University Press, 1957).

43. Baxandall and Ewen, chapter 6. Joseph Arnold, *The New Deal in the Suburbs: A History of the Greenbelt Town Program, 1935–1965* (Columbus: The Ohio State University Press, 1971). Paul K. Conkin, *Tomorrow a New World: The New Deal Community Program* (Ithaca: Cornell University Press, 1959).

44. "Houses," *Architectural Forum* 68 (June 1937): 473.

45. Baxandall and Ewen, 76–77.

46. Quoted in Kelly, *The Prefabrication of Houses,* 36.

47. Kelly, 37–38.

48. Ibid., 33.

49. Ibid., 34. See also U.S. Department of Commerce, *Building Materials and Structures Reports* (Washington, D.C.: U.S. Government Printing Office, 1939).

50. Quoted in Baxandall and Ewen, 66.

51. Kelly, *The Prefabrication of Houses,* 28–31, 38–39.

52. Ibid., 30. U.S. National Resources Committee, *Technological Trends and National Policy,* 372–73. C. Theodore Larson, "Toward a Science of Housing," *Scientific Monthly* 65 (October 1947): 295–305.

53. Ibid.

54. Walter Gropius, "Prefabrication: A Freedom from Limitations," in *Building for Modern Man, ed.* Thomas H. Creighton (Princeton: Princeton University Press, 1949), 42–48. Gilbert Herbert, *The Dream of the Factory-Made House: Walter Gropius and Konrad Wachsmann* (Cambridge, MA: MIT Press, 1984), 5–9. Howard T. Fisher, "Prefabrication: What Does It Mean to the Architect?" *Journal of the American Institute of Architects* 10 (November 1948): 219–27.

55. Miles L. Colean, "Prefabrication Needs the Architect," *Architectural Record* 90 (September 1941): 64–68. U.S. National Resources Committee, *Technological Trends and National Policy, Including the Social Implications of New Inventions* (Washington, D.C.: U.S. Government Printing Office, 1937), 370–75. Robert P. Boyce, "George Frederick Keck, 1895–1980: Midwest Architect" (Ph.D. diss., University of Wisconsin, 1986). Richard Neutra, "Prefabrication," *Arts and Architecture* 67 (June 1950): 34–44 and *Survival through Design* (New York: Oxford University Press, 1954), introduction.

56. Kelly, *The Prefabrication of Houses,* 46–48.

57. Ibid.

58. Ibid., 31–32.

59. Ibid., 32–33.

60. Ibid., 50.

61. Ibid., 38–39.

62. "Welcome to the Fordized House," *Survey* 68 (15 July 1932): 68. "Housing VI: Solutions," *Fortune* 6 (July 1932): 60–69.

63. Kelly, *The Prefabrication of Houses,* 39–40.

64. Ibid., 41. "Steel Housing Progress in 1935," *Iron Age* 137 (2 January 1936): 140–42.

65. Kelly, 41–42.

66. Ibid., 42–43.

67. "Porcelain Enamel Homes Have 27-Year History," press release by Ferro Corporation, 30 September 1959, Lustron Corporation Records, Ohio Historical Society, Box 1, Folder 7. Hereinafter cited as LCR. Robert A. Weaver, *An Ancient Art Becomes a Modern Industry: Ferro Corporation Cleveland* (New York: Newcomen Society in North America, 1956), 13–14. "Research Activities," *American Ceramic Society Bulletin* 12 (December 1933): 350.

Notes to Chapter 2

1. Greg Hise, "The Airplane and the Garden City: Regional Transformation during World War II," in *World War II and the American Dream: How Wartime Building Changed a Nation,* ed. Donald Albrecht (Cambridge, MA: MIT Press, 1995), 144–83.

2. Ibid., 151. Kristin Szylvian, "The Federal Housing Program during World War II," in *From Tenements to the Taylor Homes: In Search of an Urban Housing Policy in Twentieth-Century America, ed.* John F. Bauman, Roger Biles, and Kristin Szylvian (University Park: Pennsylvania State University Press, 2000), 120–58.

3. Szylvian, "The Federal Housing Program during World War II," 124–25.

4. Ibid. Peter S. Reed, "Enlisting Modernism," in *World War II and the American Dream: How Wartime Building Changed a Nation, ed.* Donald Albrecht (Cambridge, MA: MIT Press, 1995), 2–41.

5. Reed, "Enlisting Modernism," 4–10.

6. Ibid., 4.

7. Ibid., 11–12. By July 1943 the government expanded Lanham Act funding to $1.3 billion.

8. Ibid., 13. Kristin Szylvian, "Bauhaus on Trial: Aluminum City Terrace and Federal Defense Housing Policy during World War II," *Planning Perspectives* 9 (July 1994): 229–54. Gilbert Herbert, *The Dream of the Factory-Made House: Walter Gropius and Konrad Wachsmann* (Cambridge, MA: MIT Press, 1984).

9. Reed, "Enlisting Modernism," 18–19.

10. Ibid., 19.

11. Ibid., 12–13.

12. Szylvian, "The Federal Housing Program during World War II," 129–30. "National Housing Agency Established," *The American City* 57 (March 1942): 69.

13. Rosalyn Baxandall and Elizabeth Even, *Picture Windows: How the Suburbs Happened* (New York: Basic Books, 2000), 81–82. "The Industry Capitalism Forgot," *Fortune* 36 (August 1947): 61–67.

14. Burnham Kelly, *The Prefabrication of Houses: A Study by the Albert Farwell Bemis Foundation of the Prefabrication Industry in the United States* (Cambridge, MA: Technology Press of the Massachusetts Institute of Technology, 1951), 50.

15. Ibid., 55–63.

16. Ibid., 59. "United Front," *BusinessWeek,* 11 December 1943, 80. See also "The Magnitude

of Post-war Prefabrication," *Prefabricated Homes* 6 (November 1945): 9. Harvey Wiley Corbett, "Prefabrication in Step with Our Civilization," *Prefabricated Homes* 6 (April 1946): 13. "P.H.M.I. Expands Service to the Industry," *Prefabricated Homes* 8 (September/October 1947): 8–9.

17. Kelly, *The Prefabrication of Houses,* 60–61. National Council of American-Soviet Friends, *American-Soviet Building Conference: Proceedings* (New York: National Council of American-Soviet Friends, 1945), section 2. "Prefab Test: M.I.T. Housing Experiment Has Double Purpose—Quarters for War-Veteran Students and Research in Prefabrication," *Business Week,* 17 November 1945, 44.

18. "P.H.M.I. Expands Service to the Industry," *Prefabricated Homes* 8 (September/October 1947): 8–9. For examples of thoughtful criticism of prefabrication, see Joseph Hudnut, "The Post-Modern House," *Architectural Record* 97 (May 1945): 70, and Albert Charles Schweizer, "Factory-Built Houses an Urban Threat: But Foresight May Bring Benefits from Prefabrication," *The American City* 58 (December 1943): 64–65.

19. Kelly, *The Prefabrication of Houses,* 62–63.

20. Baxandall and Ewen, *Picture Windows,* 87.

21. Richard O. Davies, *Housing Reform during the Truman Administration* (Columbia: University of Missouri Press, 1966), introduction. See also Susan Hartmann, *Truman and the 80th Congress* (Columbia: University of Missouri Press, 1971), and Barton J. Bernstein, "Reluctance and Resistance: Wilson Wyatt and Veterans' Housing in the Truman Administration," *Register of the Kentucky Historical Society* 65 (January 1967): 44–66. In a March 1946 public opinion poll, twenty-seven percent of Americans responded that they were affected by the housing shortage. Among veterans, forty-two percent indicated a hardship resulting from the housing shortage. These figures appear in *Public Opinion Quarterly* 10 (Summer 1946): 260.

22. Gwendolyn Wright, *Building the Dream: A Social History of Housing in America* (New York: Pantheon Books, 1981), 242. Mary Frost Jessup, "Trends in Housing during the War and Postwar Periods," *Monthly Labor Review* 64 (January 1947): 11–14. Marjorie Foulkrod, "The Case History of Housing," *Current History* 15 (September 1948): 138–40. Charles Abrams, "Homeless America," *Nation* 163 (21 and 28 December 1946): 723–25, 753–55. "Mr. Wyatt's Shortage," *Fortune* 33 (April 1946): 105.

23. Bernstein, "Reluctance and Resistance," 47–48.

24. Ibid., 49–50.

25. Ibid., 51–53.

26. Executive Order 9686, 11 *Federal Record* 1033. See "Mayor Wyatt of Louisville," *Fortune* 30 (November 1944): 139–45.

27. Truman to Wyatt, 12 December 1946, Harry S. Truman Papers, Harry S. Truman Library, Independence, MO, Official File 63. Hereinafter cited as Truman Papers. Wilson W. Wyatt, *Whistle Stops: Adventures in Public Life* (Lexington: University Press of Kentucky, 1985), 59.

28. Office of the Housing Expediter, press release, 5 January 1946 (Office of the Housing Expediter Records, Record Group 252, National Archives, Washington, DC). Hereinafter cited as Housing Expediter Records.

29. Wyatt, *Whistle Stops,* 61–62. Coleman Woodbury, "Objectives and Accomplishments of the Veterans Emergency Housing Program," *The American Economic Review* 37 (May 1947): 508–23.

30. Wilson W. Wyatt, *Veterans' Emergency Housing Program, Report to the President, 7 February 1946* (Washington, D.C.: U.S. Government Printing Office, 1946), i.

31. Wyatt, *Veterans' Emergency Housing Program,* 3.

32. Ibid., 1–3.

33. Ibid, 2, 7–8. The Veterans' Emergency Housing Act as approved by Congress on May 22, 1946 provided RFC backing for prefabrication only "temporarily until they attain general market acceptability" and only if prefab companies caused no "economic dislocations or adverse effects upon established businesses." *United States Statutes at Large* 60 (Part I), 215.

34. Woodbury, "Objectives and Accomplishments of the Veterans Emergency Housing Program," 511.

35. *Congressional Record,* 79th Cong., 2d sess., 1 March 1946, 1829–48; 4 March 1946, 1863–91; 5 March 1946, 1940–64; and 6 March 1946, 1974–95.

36. Wyatt, *Whistle Stops,* 70. *United States Statutes at Large* 60 (Part I), 207–15.

37. Office of the Housing Expediter, press release, 22 May 1946, Housing Expediter Records.

38. *Time* 47 (8 April 1946): 22–23. Wyatt, *Whistle Stops,* 71.

39. "The Promise of the Shortage," *Fortune* 33 (April 1946): 101–103. Foulkrod, "Case History of Housing." Robert Lasch, *Breaking the Building Blockade* (Chicago: University of Chicago Press, 1946) and Dorothy Rosenman, *A Million Homes a Year* (New York, 1945) offer discussions of the vagaries and problems of the residential construction industry in the period immediately following World War II. "Mr. Wyatt Builds His Castle," *Fortune* 24 (July, 1946): 3.

40. U.S. National Housing Agency, *Report of the Veterans' Emergency Housing Program* 1 (July, August, and September 1946).

41. Davies, *Housing Reform,* 51.

42. Alfred P. Sloan, *Adventures of a White Collar Man* (New York: Doubleday, Doran & Company, 1941), 197–98. Wyatt, *Whistle Stops,* 72.

43. *Lustron: A New Standard for Living,* documentary film by Ed Moore and Bill Kubota, October, 2001. Rough cut in possession of the author.

44. *Minneapolis Tribune,* 12 September 1982.

45. Ibid. Thomas T. Fetters and Vincent Kohler, *Lustron Homes: The History of a Postwar Prefabricated Housing Experiment* (Jefferson, NC: McFarland, 2002), 10–12.

46. Ibid.

47. "President's Report—1942," Presidents File, Chi-Vit Archive, Cabinet 1, Folder 4. Herein after cited as CVA. 1–2. U.S. Congress, Senate, Committee on Banking and Currency, *Inquiry into the Operation of the Reconstruction Finance Corporation and Its Subsidiaries under Senate Resolution 132,* pt. 2, 80th Cong., 2d sess., 14–16 and 22 January 1948, 343. Harry J. VanDolah interview, 28 September 1985.

48. *Inquiry into the Operation of the Reconstruction Finance Corporation,* 348–49. Arthur Bartlett, "Lustron—The House That Lots of Jack Built," *Collier's* 124 (5 November 1949): 15.

49. *Congressional Record,* 12 January 1951, 193–95. Strandlund's "press quench" method solved the problem of distortion in armor plating during the tempering process. Rather than immersing the hot plates in a bath of cold water, which caused warping, the press quench method cooled the steel in high-pressure presses with thousands of water injectors spread evenly across the surface. This produced a ready-to-use, perfectly flat plate that did not have to be cold hammered to compensate for distortions in the production process.

50. "Lusterlite Laboratories," Advertising File, CVA, Cabinet 1, Folder 2. "Research Laboratories Finished in Vitreous Enamel," *American Ceramic Society Bulletin* 14 (December 1935): 404–6. "The Laboratory Goes Modern," *Chemical Industries* 74 (December 1935): 575.

51. William Hogenson, president of the Chicago Vitreous Enamel Products Company, "Industrial Origins of Porcelain Enameling," speech before the Second Annual Conference of the Porcelain Enamel Institute, Chicago, IL, April 1932, Presidents File, CVA, Cabinet 1, Folder 1. Harry J. VanDolah, former president of the Chi-Vit Corporation, interview with author, Palm Beach, FL, 28 September 1985. "Modern Wonders of an Ancient Art," Pamphlet File, CVA, Cabinet 1, Folder 3.

52. Ibid. "Production at Chicago Vitreous," unsigned paper, Miscellaneous File, CVA, Cabinet 1, Folder 4.

53. Hogenson, "Industrial Origins," 2. Robert A. Weaver, *An Ancient Art Becomes a Modern Industry: Ferro Corporation Cleveland* (New York: Newcomen Society in North America, 1956), 17.

54. Winfred S. Downs, ed., *Encyclopedia of American Biography* (New York, 1939), 1. "William Hogenson," American Historical Society, Presidents File, CVA, Cabinet 1, Folder 2. "The Laboratory Goes Modern," 575.

55. "P.E.I. Founders Honored at 26th Annual Meeting," *American Ceramic Society Bulletin* 37 (February 1958): 106. William Hogenson to Mr. Walters, 30 September 1931, Presidents File, CVA, Cabinet 1, Folder 3. "Our Industry," *Better Enameling* 1 (January 1930): 3.

56. See, for example, "Porcelain Enamel Will Make the All-Steel House Practical," *Better Enameling* 2 (March, 1931): 4–9. "A New Type of Construction for Porcelain Enameled Buildings," *Better Enameling* 3 (August, 1932): 4–6, 36, 40. "A Frameless Steel House with Porcelain Enameled Exterior," *Better Enameling* 3 (September, 1932): 21.

57. "A Dual Purpose Laboratory," *Better Enameling* 7 (January 1936): 7. "Lusterlite Laboratories," 2. The importance of industrial research facilities and their impact on their parent companies, industry competition, and industrial structure is discussed in Leonard S. Reich, *The Making of American Industrial Research: Science and Business at GE and Bell, 1876–1926* (New York: Cambridge University Press, 1985). Research laboratories in the enameling industry followed the pattern of development described by Reich, i.e., that industrial research was concentrated in areas that were most technologically complicated and most promising for commercial application. Lusterlite and Ferro Corporation research laboratories were fundamentally scientific research centers, yet Lusterlite is unusual for its emphasis on sales and promotion. It produced numerous informational brochures and colorful pamphlets for distribution to architectural firms and commercial building contractors touting the beauty, indestructibility, and design flexibility of enameled-steel panels. Lusterlite also ran an informal "speakers bureau" to promote the company's research efforts among the general public.

58. "Chicago Vit in the News," unsigned memorandum, Advertising File, CVA, Cabinet 1, Folder 2.

59. "Lusterlite Laboratories," 3.

60. See Norman Bel Geddes, *Horizons* (Boston: Little, Brown, and Company, 1932); Walter Dorwin Teague, *Design this Day: The Technique of Order in the Machine Age* (New York: Harcourt, Brace, 1940); and Raymond Loewy, *Industrial Design* (Woodstock, NY: Overlook Press, 1979), for first hand accounts of the "movement."

61. "President's Report—1935," Presidents File, CVA, Cabinet 1, Folder 4.

62. Emanuel Hogenson to William Bryant, 4 April 1937, Secretarial File, CVA, Cabinet 1, Folder 2. "The Factory Built House Is Here, But Not the Answer to the $33 Million Question: How to Get It to Market?" *Architectural Forum* 90 (May 1949): 109. R. Harold Denton to Gerald Fritz, 10 April 1967, Lustron Corporation Records, Ohio Historical Society, Columbus, OH, Box 1, Folder 4. Hereinafter cited as LCR.

63. "Chi-Vit Heritage," unsigned promotional paper ca. 1948, Advertising File, CVA, Cabinet 1, Folder 3.

64. Harry J. VanDolah interview, 28 September 1988. James L. Pease to Robert A. Irwin, 15 October 1946, LCR, Box 1, Folder 4.

65. Articles of Incorporation, Porcelain Products Company, 15 July 1937, PPC File, CVA, Cabinet 1, Folder 5.

66. Burnham Kelly, *The Prefabrication of Houses: A Study by the Albert Farwell Bemis Foundation of the Prefabrication Industry in the United States* (Cambridge, MA: Technology Press of the Massachusetts of Technology, 1951), 29, 38–39.

67. James L. Pease to Robert A. Irwin, 2. "Chicago Vitreous Demonstrates Improved Architectural Porcelain Enamels," *American Ceramic Society Bulletin* 19 (May 1940): 203.

68. Ibid.

69. *Official Gazette of the U.S. Patent Office* 595 (February 1947): 440.

70. *Inquiry into the Operation of the Reconstruction Finance Corporation,* 344–45.

71. Strandlund to William Hogenson, 25 June 1946, President's File, CVA, Cabinet 1, Folder 4.

72. Ibid. *Inquiry into the Operation of the Reconstruction Finance Corporation,* 349–50. Bartlett, "Lustron—The House That Lots of Jack Built," 15–16.

73. Fetters and Kohler, *The Lustron Home,* 16–17.

74. R. Harold Denton to Harold D. Hauf, 13 August 1946, LCR, Box 1, Folder 4.

75. "Strandlund June–July Misc.," collection of loose notes and telephone logs for William Hogenson, President's File, CVA, Cabinet 1, Folder 4. Stephen B. Adams, "Mr. Kaiser Goes to Washington: Saga of a Government Entrepreneur" (Ph.D. diss., Johns Hopkins University, 1994), introduction.

76. Strandlund to William Hogenson, 12 July 1946, President's File, CVA, Cabinet 1, Folder 4. *Official Gazette of the U.S. Patent Office* (28 December 1937): 793.

77. Roy B. Blass, "The 'Lustron' Plan for Designed Housing: A Conception of a Housebuilding Program on a Mass Production Basis," Miscellaneous Files, CVA, Cabinet 2, Folder 2. "The Industrialized House: Lustron Develops an Enameled Steel House," *Architectural Forum* 86 (June 1947): 105–10.

78. Ibid. See also Gwendolyn Wright, *Building the Dream: A Social History of Housing in America* (New York: Pantheon Books, 1981), 253–54; and Elaine Tyler May, *Homeward Bound: American Families in the Cold War Era* (New York: Basic Books, 1988), 58–91.

79. Barbara L. Allen, "The Ranch-Style House in America: A Cultural and Environmental Discourse," *Journal of Architectural Education* 49, 3 (February 1996): 156–65.

80. Ibid. Clifford Edward Clark, Jr., *The American Family Home, 1800–1960* (Chapel Hill: University of North Carolina Press, 1986), 193–216.

81. Blass, "The 'Lustron' Plan," 1–7.

82. Ibid. "Lustron: A Prefabricated Ranch House of Porcelainized Steel," *Fine Homebuilding* (August/September 1984): 26–29.

83. Ibid.

84. Ibid.

85. Ibid.

86. R. Harold Denton to Robert A. Irwin, 4 October 1946, LCR, Box 1, Folder 4.

87. Ibid., 2.

88. Ibid., 3.

89. Ibid., 4. National Housing Administration, Office of the National Housing Expediter, "Report to the Kilgore Committee," LCR, Box 1, Folder 13.

90. "Report to the Kilgore Committee," 1.

91. U.S. Congress, Senate, *Investigation of the National Defense Program: Hearings before a Special Committee Investigating the National Defense Program,* 79th Cong., 2d sess., 5–7, 9, 26–27, 30 September 1946; 20–21, 23, 25 November 1946; and 31 January 1947, 23027–42.

92. R. Harold Denton to Robert A. Irwin, 4 October 1946, LCR, Box 1, Folder 4.

93. Ibid., 2–3.

94. Ibid., 3–4. *Inquiry into the Operation of the Reconstruction Finance Corporation,* 348–60.

95. Ibid. In a letter to Truman, Senator Albert Hawkes expressed serious concern that the government would even consider lending $52 million to a concern "whose owners have only put up $36,000, and thus put them in preferred competition with established businesses." Hawkes asked, "[W]hat hope there is for preserving the free enterprise system or legitimately protecting established business in this country[?]" Hawkes to Truman, 23 November 1946, Truman Papers, Official File 1930.

96. James L. Pease to Robert A. Irwin, 15 October 1946, LCR, Box 1, Folder 4.

97. Ibid., 2–3. Chicago Vitreous's innovation of low-temperature curing enamels, which prevented heat distortion in the metal, provided the key to mass production. The use of high-speed forming rolls, fast acting presses, sophisticated material-handling systems, and continuous-process enameling furnaces assured that Lustron would be on the cutting edge of manufacturing technology. Many of these production innovations had been developed during the war in a variety of industries.

98. Ibid.

99. Ibid., 3–4.

100. Ibid.

101. Ibid.

102. Omar B. Ketchum to George E. Allen, 29 October 1946, LCR, Box 1, Folder 4.

103. "Wyatt Asks RFC, WAA for House-Making Aid," *New York Times,* 24 October 1946, 33. "Wyatt Asks Loans to Aid Homes," *New York Times,* 26 October 1946, 7. In addition to the $52 million loan request for Lustron, Wyatt sought $40 million in additional loans for prefabricated housing ventures by the Higgins Company, Douglass Aircraft Corporation, and Consolidated Vultee Aircraft Company. "Where Is Prefabrication?" *Fortune* 33 (April 1946): 127–31.

104. Charles B. Henderson to Wilson Wyatt, 31 October 1946, LCR, Box 1, Folder 4. "RFC Blocks Loans to Speed Housing," *New York Times,* 2 November 1946, 32. Allen's opposition stemmed in part from his promise to Truman to keep the RFC "scandal-free." See George E. Allen, *Presidents Who Have Known Me* (New York: Simon and Schuster, 1950), 207–9.

105. Henderson to Wyatt, 31 October 1946, 2. "Wyatt Wins, Loses on Housing Loans," *New York Times,* 26 November 1946, 23.

106. Wilson W. Wyatt to Charles B. Henderson, 1 November 1946, LCR, Box 1, Folder 4.

107. Ibid. Herbert P. Snyder to Wilson W. Wyatt, 1 November 1946, LCR, Box 1, Folder 4.

108. Wyatt to Henderson, 2.

109. Veterans of Foreign Wars of the United States, Press Relations Department, 13 November 1946, LCR, Box 1, Folder 4.

110. Louis E. Starr to President Truman, 13 November 1946, LCR, Box 1, Folder 4.

111. "Prefabs: Wyatt's Choice," *Newsweek* 28 (2 December 1946): 76. Lustron

Scrapbook, Miscellaneous Files, CVA, Cabinet 2, Folder 3. Robert A. Mitchell, "What Ever Happened to Lustron Homes?" *APT Bulletin* 23 (1991): 46.

112. Carl Strandlund to Senator Homer Ferguson, 22 November 1946, LCR, Box 1, Folder 4. Lustron Scrapbook, Miscellaneous Files, CVA, Cabinet 2, Folder 3. "Prefabs: Wyatt's Choice," *Newsweek* 28 (2 December 1946): 76.

113. Wyatt's power to secure plants for housing production stemmed from Sections 2 (b) (2) and 4 (a) of the Veterans' Emergency Housing Act of 1946 (Public Law 388, 79th Cong.), from Executive Order 9686, and from the allocation authority delegated to Wyatt in his capacity as Housing Expediter by Directive 44 of the Civilian Production Administration (11 *Federal Record* 8396) under the Second War Powers Act of 1942.

114. "Report to the Kilgore Committee," 10–11. The WAA issued on September 18 a "letter of intent" to Tucker, promising to deliver the plant "upon submission of evidence by Tucker Corporation that said Corporation has to its credit, as the result of public or private financing, the sum of $15 million in cash by March 1, 1947."

115. "Statement by the President Terminating Price and Wage Controls," *Public Papers of the Presidents of the United States: Harry S. Truman, 1946* (Washington, D.C.: U.S. Government Printing Office, 1962), 475–77. Truman's promise to end controls as soon as possible can be found in his "Message on Reconversion," 6 September 1945. John D. Small, head of the Civilian Production Administration and a close Truman advisor, emerged as one of Wyatt's most vocal critics. His conservative economic philosophy conflicted with the reliance of the VEHP on governmental controls. Small also disliked Wyatt personally and believed that ambition had clouded his judgment. Small complained to White House aide Clark Clifford that Wyatt equated criticism of the VEHP with "opposition to veterans' housing." Quoted in Richard O. Davies, *Housing Reform during the Truman Administration* (Columbia: University of Missouri Press, 1966), 52.

116. Wyatt's comment is quoted in Nathaniel S. Keith, *Politics and the Housing Crisis since 1930* (New York, 1973), 65. The Truman quotation is from the "Statement by the President Terminating Wage and Price Controls," *Public Papers of the Presidents: Harry S. Truman,* 476.

117. U.S. War Mobilization and Reconversion Office, *Address by Wilson W. Wyatt, Housing Expediter and Administrator, National Housing Agency, over Mutual Broadcasting Company, 14 November 1946.* Davies, *Housing Reform,* 56–57.

118. "Statement of Wilson W. Wyatt, National Housing Expediter, before Senate Committee to Investigate the National Defense Program," 25 November 1946, LCR, Box 1, Folder 4.

119. Ibid., 4–6.

120. "Report of the Housing Expediter for 1947,"Truman Papers, Official File 63. Wilson W. Wyatt to President Truman, "Memoranda,"Truman Papers, Official File 122-I. "Wyatt v. Everybody," *Time* 48 (25 November 1946): 89–90. "A Huff and a Puff," *Time* 48 (9 December 1946): 24.

121. Wilson W. Wyatt to President Truman, 4 December 1946, Truman Papers, Official File 63. *Public Papers of the Presidents: Harry S. Truman,* 488–90. Wyatt, *Whistle Stops,* 85–86. R. Harold Denton resigned from the NHA on November 5. Lauding Wyatt, he blamed the "real estate and builders' lobbies" for blocking government support for mass-produced housing: "It is a clear-cut fight against progress by a backward industry—an industry whose thinking is geared to low volume, large profits, poor quality, insufficient methods, and greed." R. Harold Denton to Wilson W. Wyatt, and Wyatt to Denton, both 5 December 1946, LCR, Box 1, Folder 4. In 1947 Denton joined Lustron as Director of Market Development.

122. Carl Strandlund, "For Release December 6, 1946," LCR, Box 1, Folder 4.

123. Ibid., 2. "Prefab Firm Drops Interest in Dodge Plant," *The Washington Post,* 6 January 1947, 7. "Scaledown for Lustron," *Business Week,* 1 February 1947, 16.

124. *Inquiry into the Operation of the Reconstruction Finance Corporation,* 351–52. Bartlett, "Lustron—The House That Lots of Jack Built," 69. "That Lustron Affair," *Fortune* 40 (November 1949): 93

125. *Inquiry into the Operation of the Reconstruction Finance Corporation,* 351–55. "RFC Approves Loan on Lustron Home," *New York Times,* 31 January 1947, 35. "Scaledown for Lustron," *Business Week,* 1 February 1947, 16.

126. Ibid.

127. Ibid.

Notes to Chapter 3

1. Historian Richard O. Davies referred to 1947 as the "year of postponement" as continuing political furor over the Taft-Ellender-Wagner housing bill, especially its provisions for extending funding for public housing, confounded the Truman Administration's efforts to adopt a "national housing policy affirming the government's responsibility in postwar housing." The administration's support for Lustron fit Truman's desire for productive reform in the housing industry, best expressed by Raymond Foley (Wyatt's successor at the NHA) who called for "determined action by private enterprise to reduce costs and provide houses that more Americans of moderate and low incomes can afford." Indeed, using government capital to seed innovative enterprises like Lustron suited the administration's desire to drive productive solutions in an effort to reduce the need and thus the political controversies surrounding public housing. Davies, *Housing Reform during the Truman Administration* (Columbia: University of Missouri Press, 1966), 32, 61.

2. U.S. Congress, Senate, Committee on Banking and Currency, *Inquiry into the Operation of the Reconstruction Finance Corporation and Its Subsidiaries under Senate Resolution 132,* pt. 2, 80th Cong., 2d sess., 14–16 and 22 January 1948, 350–54. Arthur Bartlett, "Lustron—The House That Lots of Jack Built," *Collier's* 124 (5 November 1949): 69–70.

3. "Housing Experts Discuss Problem," *New York Times,* 17 April 1947, 31. Prins had also tried to get Strandlund in to see President Truman in February, but the president refused. "Mr. Carl G. Strandlund," 15 February 1947, Truman Papers, Official File 1930.

4. Ibid.

5. U.S. Congress, House of Representatives, Committee on Banking and Currency, *Housing and Rent Control,* 80th Cong., 1st sess., 17–21 and 24–28 March 1947, 1–2. *United States Statutes at Large* 60 (Part I): 207–15.

6. Burnham Kelly, *The Prefabrication of Houses: A Study by the Albert Farwell Bemis Foundation of the Prefabrication Industry in the United States* (Cambridge, MA: Technology Press of the Massachusetts Institute of Technology, 1951), 172–73. *Housing and Rent Control,* 486–87.

7. *Housing and Rent Control,* 486.

8. Ibid., 487.

9. Ibid., 488.

10. Ibid., 490.

11. Ibid., 491–92. "The Industrialized House: Lustron Corp. Develops an Enameled Steel House," *Architectural Forum* 86 (June 1947): 105–10.

12. Ibid., 494–95.

13. John R. Steelman to John D. Goodloe, 30 June 1947, Truman Papers, Official File 1930. U.S. Congress, Senate, Committee on Banking and Currency, *Inquiry into the Operation of the Reconstruction Finance Corporation and Its Subsidiaries under Senate Resolution 132,* pt.1, 80th Cong., 1st sess., 3, 11–12 December 1947, 235–36. U.S. Housing and Home Finance Agency, *Chronology of Major Actions Affecting Housing and Community Development, 1892–1963* (Washington, D.C.: U.S. Government Printing Office, 1964), 12.

14. Steelman to Goodloe, 30 June 1947. *Inquiry into the Operation of the Reconstruction Finance Corporation,* pt.1, 236.

15. Goodloe to Senator C. Douglass Buck, 7 July 1947, Truman Papers, Official File 1930.

16. Board of Directors of Reconstruction Finance Corporation, "Resolution," 30 June 1947, LCR, Box 1, Folder 13. "Lustron Corporation, Cicero, IL," 30 June 1947, Truman Papers, Official File 1930. For its guarantee, Chicago Vitreous received one share of Lustron common stock for each $10 expended on development work.

17. Ibid. Lustron was also required to apply the proceeds from the sale of additional capital stock to the repayment of the loan.

18. "$15,500,000 Housing Loan," *New York Times,* 6 July 1947, 42.

19. *Inquiry into the Operation of the Reconstruction Finance Corporation,* pt. 2, 371.

20. Ibid., 345–47. Board of Directors of Reconstruction Finance Corporation, "Resolution," 30 June 1947. E. Merl Young to William Willett, "Memorandum to Mr. Willett," 7 August 1947, Truman Papers, Official File 1930.

21. Emanuel Hogenson to James Conger, 12 July 1947, Correspondence File, CVA, Cabinet 1, Folder 10.

22. *Inquiry into the Operation of the Reconstruction Finance Corporation,* pt. 2, 347. Strandlund to RFC, 7 August 1947, Truman Papers, Official File 1930.

23. Strandlund to RFC, 7 August 1947.

24. *Inquiry into the Operation of the Reconstruction Finance Corporation,* pt. 2, 347, 351–52.

25. Ibid., 530–31.

26. "Concern Formed to Build Houses," *New York Times,* 1 November 1947, 22. W. V. Burnell to W. E. Willett, 15 July 1948, Truman Papers, Official File 1930.

27. *Inquiry into the Operation of the Reconstruction Finance Corporation,* pt. 2, 529–32, 535–40.

28. Ibid, 526. Burnell to Willett, 15 July 1948.

29. Ibid.

30. "AFL in Pact Covering Ready-Made Houses," *New York Times,* 12 November 1947, 48. "Agreement between Lustron Corporation and the United Brotherhood of Carpenters and Joiners of America," 15 April 1947, Carpenters Local 1138 Collection, Center for Archival Collections, Bowling Green State University, MS 147, Box 8, Folder 2. Hereinafter cited as Carpenters Local Collection.

31. Ibid. R. Harold Denton to Gerald Fritz, 10 April 1967, LCR, Box 1, Folder 4. *Housing and Rent Control,* 498–99. *Inquiry into the Operation of the Reconstruction Finance Corporation,* pt. 2, 367. "Is Labor to Blame?" *Business Week,* 10 January 1948, 86–87. *Lustron Newsletter,* 12 March 1948, in possession of the author.

32. *Lustron Newsletter,* 15 February 1948.

33. "The Lustron Loan," *Ohio State Journal,* 20 December 1947, 12. Descriptions of specific industrial equipment for the plant can be found in the following sources: "Machinery for Lustron House," *New York Times,* 5 December 1947, 45; "Lustron Signs Contract for

Million Dollar Oven," *Columbus Dispatch,* 27 December 1947; and "Lustron Corp. Places Orders for Equipment," *Columbus Dispatch,* 11 February 1948.

34. Andrew J. Dunar, *The Truman Scandals and the Politics of Morality* (Columbia: University of Missouri Press, 1984), 78–79.

35. *Inquiry into the Operation of the Reconstruction Finance Corporation,* pt. 2, 362–72.

36. Ibid., 363, 373. "Head of Lustron Denies 'Pressure,'" *Ohio State Journal,* 16 January 1948. "Lustron Head Reveals Only $1000 Investment," *Columbus Dispatch,* 16 January 1948.

37. *Inquiry into the Operation of the Reconstruction Finance Corporation,* pt. 2, 402–5. "RFC Lustron Loan Called Unsound," *New York Times,* 12 December 1947.

38. Ibid.

39. *Inquiry into the Operation of the Reconstruction Finance Corporation,* pt. 2, 410.

40. Joint Committee on Housing of the 80th Congress, *High Cost of Housing* (Washington, 1948). "Lustron-Made House Low in Cost, House Told," *Columbus Dispatch,* 20 January 1948. House Concurrent Resolution 104, 80th Cong., 26 July 1947, authorized the creation of a Joint Committee on Housing to "study all phases of housing." The Committee conducted hearings in Washington, D.C., and thirty-two other cities to gauge housing problems and elicit advice and opinions on pending and future legislation. The Committee was essentially a smokescreen on the part of the Republican majority to cover its inaction on the Taft-Ellender-Wagner bill and to provide a forum in which to criticize public housing. Its positive views on government–industry cooperation to stimulate housing production created an interesting opening for Lustron, however, and provided some political shelter and rationalization for continued government investment.

41. Quoted in Rosalyn Baxandall and Elizabeth Ewen, *Picture Windows: How the Suburbs Happened* (New York: Basic Books, 2000), 93. McCarthy took an interest in Lustron in part because it fit his stated position on subsidizing the development of housing technologies and firms. Perhaps more intriguing to McCarthy was that such support held the promise of financial gain. Lustron was anxious to exploit political connections, and McCarthy was anxious to boost his image as a player in housing politics. The result was the publication "Wanted: A Dollar's Worth of Housing for Every Dollar Spent," originated by and credited to McCarthy but polished by several others into a 96-page pamphlet promoting prefabrication in general and Lustron in particular. Strandlund later testified before a congressional committee investigating McCarthy that the company had paid McCarthy $10,000 for the "article," which had to be rewritten at an additional expense of $2,000 before McCarthy could release it publicly. It was a quid pro quo, pure and simple, and at $1.43 per word, it proved one of the more expensive purchases the company made.

42. *Lustron Newsletter,* 5 March 1948.

43. "Lustron's First Enameled Steel Home to Be Unveiled Next Month," *Columbus Dispatch,* 29 February 1948.

44. "Lustron Appoints New York Agents," *New York Times,* 4 April 1948, 7. "First House Built by Lustron to Be Displayed in N.Y.," *Columbus Dispatch,* 2 April 1948. *Lustron Newsletter,* 2 April 1948.

45. Ibid. A Lustron advertisement that appeared in *Architectural Forum* 90 (June 1949): 4, used a picture of the crowds attending the New York City model home. "Bulletin of Progress," LCR, Box 1, Folder 2. *Lustron Newsletter,* 16 April 1948.

46. "Site Plan Showing Planting Scheme for Lustron Demonstration Home New York City," reproduced in Thomas T. Fetters and Vincent Kohler, *Lustron Homes: The History of a Postwar Prefabricated Housing Experiment* (Jefferson, NC: McFarland, 2002), 46.

47. *Lustron Newsletter,* 16 and 23 April 1948.

48. *Lustron Newsletter,* 30 April and 7 May 1948.

49. "Lustron House Is Displayed in Washington," *Columbus Dispatch,* 29 May 1948. "Porcelain Enameled Home to Be Opened to Public," *Washington Post,* 27 May 1948. "Housing Still Top Problem, VFW Head Asserts," *Columbus Dispatch,* 11 February 1948. *Lustron Newsletter,* 28 May 1948.

50. "Lustron Home to Be on Display Saturday," *Columbus Dispatch,* 8 November 1948.

51. "Lustron Holds Preview to Unveil New All-Porcelain Home," *Columbus Dispatch,* 11 November 1948.

52. *Lustron Newsletter,* 28 May 1948. The AFL recruited Peterson's assistant, Phil Delaney, from Lustron in July 1948 to serve as one of three AFL representatives to assist with labor issues for the European Cooperative Administration.

53. Miscellaneous Advertisements, LCR, Box 1, Folder 1; Box 2, Folders 1, 2, and 3. "Lustron's Mat Advertising Service," LCR, Box 2, Folder 1. The "Mat" service provided advertising templates to dealers, who then inserted their names, for local and regional distribution. Good examples of Lustron advertisements can be found in *Life,* 13 December 1948, 118; and *Time,* 20 September 1948, 69 and 17 January 1949, 61. "Bulletin of Progress," LCR, Box 1, Folder 2. Lustron employed the New York advertising agency Leo Burnett as its primary national agency. James V. Barton was Lustron's account executive.

Lustron's promotional expenditures aroused some concern on the part of Housing Expediter Tighe Woods, who wrote to Steelman in December 1948 suggesting that Strandlund "go easy on the promotion and advertising budget because if there is any product in America today that needs less advertising it's the Lustron house. Their dealers have had to go into hiding to keep the public from doing them bodily harm." Woods to Steelman, 4 December 1948, Truman Papers, Official File 1930.

54. Fetters and Kohler, *Lustron Homes,* 47.

55. *Lustron Newsletter,* 5 March 1948.

56. Ibid., 18 June 1948. Fetters and Kohler, *Lustron Homes,* 27–28.

57. "The Home America Has Been Waiting For," LCR, Box 1, Folder 1. "The Lustron Home: A New Standard for Living," LCR, Box 1, Folder 1.

58. "Threshold to Better Living," LCR, Box 2, Folder 1.

59. "The Home America Has Been Waiting For," LCR, Box 1, Folder 1.

60. "Threshold to Better Living," LCR, Box 2, Folder 1. "Lustron Gives a Man a Home He Can Own," LCR, Box 2, Folder 1.

61. "Threshold to Better Living," LCR, Box 2, Folder 1. "Tomorrow's Home . . . Today," LCR, Box 2, Folder 3. "Some Things You'll Want to Know about Lustron's Porcelain Enamel," LCR, Box 2, Folder 1. Lustron continued to emphasize the "health issue" and eventually obtained the endorsement of the Mayo Clinic, which "recommended on more than one occasion to sufferers of the dreaded disease known as asthma that they live in a Lustron house." Quoted in *Congressional Record,* 8 February 1951, A666.

62. "Answers to Your Questions about the Lustron Home," LCR, Box 1, Folder 1. See also, "Basements Not Needed in Homes Today, Lustron Executive Says," *Columbus Dispatch,* 5 August 1948.

63. "Lustron Sales Program," LCR, Box 1, Folder 9.

64. Ibid.

65. "Porcelain Prefab Nearly Ready," *Business Week,* 24 April 1948, 39–44. "Lustron Newsletter," 4 June 1948 and 8 October 1948, LCR, Box 1, Folder 13.

66. R. Herald Denton to Gerald Fritz, 10 April 1967, LCR, Box 1, Folder 4. "Truck

Is Package for House," *New York Times,* 20 October 1948, 43. E. J. Clement, "Assembly Line to Home Site," *Mill and Factory* 45 (September 1949): 117–21.

67. Fetters and Kohler, *Lustron Homes,* 63–65.

68. Ibid.

69. *Lustron Newsletter,* 14 May 1948.

70. Ibid., 7 February 1948. A poll conducted by the company in August 1948 found that 50 percent of the employees did not own their own homes, but 128 out of 145 respondents expressed a desire to own a Lustron home. Also, the poll revealed that nearly 50 percent of Lustron employees had served in the armed forces during the war. In October 1948 one of Lustron's attorneys, John W. Wilke, offered employees the opportunity to purchase lots he owned for $879 and the company agreed to consider allocating a number of early production houses as models for other potential Lustron neighborhoods.

71. Ibid., 19 February 1948.

72. "Steel and Housing," *Iron Age,* 26 January 1950, 66–68. Ernest Olsen, "Handling and Fabricating Methods for Producing the Lustron Home," *Finish,* October 1948, 9–12. Olsen was hired as Lustron's general superintendent of production in April 1948.

73. Ibid.

74. "Lustron's Spot-Welded Homes," *The Welding Engineer* 34 (July 1949): 17–23.

75. "First in New Ceramic Era: Lustron's $13 Million Plant," *Ceramics Industry* 51 (August 1948): 56–58. "Nation Watches Lustron Mass Produce Housing," *Ohio State Journal,* 27 August 1948. The enameling process, adapted from technology developed by Chicago Vitreous, was one of Lustron's most important technological advantages. The low-temperature firing of porcelain enamel could now be accomplished in one coat, instead of applying a base coat and a finish coat. The process also permitted the use of ordinary steel rather than more expensive enameling iron, since lower temperatures eliminated warping that would result from conventional enameling of ordinary steel. Chicago Vitreous was the recipient of Lustron's initial order for $1,206,445 of enameling compounds—the largest single order in the history of the Hogenson's enterprise.

76. "First in New Ceramic Era," 56–57. "Layout of Porcelain Enameling Department, Lustron Corp.," *Ceramic Industry* 51 (September 1948): 68–70.

77. Olsen, "Handling and Fabricating Methods for Producing the Lustron Home," 11–12. "Lustron's $14,000,000 Gamble," *Engineering News Record* 141 (25 November 1948): 60–63.

78. "That Lustron Affair," *Fortune* 40 (November 1949): 93–94.

79. Kelly, *The Prefabrication of Houses,* 303–4. Carl Koch and Andy Lewis, *At Home with Tomorrow* (New York: Rinehart, 1958), 62, 126.

80. "The Factory Built House Is Here, But Not the Answer to the $33 Million Question: How to Get It to Market," *Architectural Forum* 90 (May 1949): 107.

81. "Bathtub Blues," *Time* 54 (4 July 1949): 55.

82. Kelly, *The Prefabrication of Houses,* 80. Koch and Lewis, *At Home with Tomorrow,* 112. Richard Bender, *A Crack in the Rear-View Mirror: View of Industrialized Building* (New York: Van Nostrand Reinhold, 1973), 60.

83. U.S. Housing and Home Finance Agency, *Chronology of Major Actions Affecting Housing and Community Development, 1892–1963* (Washington, DC, 1964), 15.

84. "Re: Lustron Corporation," n.d., Truman Papers, Official File 1930.

85. W. V. Burnell to W. E. Willett, 15 July 1948, Truman Papers, Official File 1930.

86. Ibid.

87. Ibid. W. V. Burnell to W. E. Willett, 8 July 1948, Truman Papers, Official File 1930.

88. "RFC Grants Lustron New $10 Million Loan," *Columbus Dispatch,* 21 July 1948.

"Lustron Will Get $10,000,000 More," *New York Times,* 22 July 1948, 40. "The Winner: First Industrialized House Looks like a Sure Bet As Lustron Clinches Loan," *Architectural Forum* 87 (August 1948): 14. "Facts about Lustron," LCR, Box 1, Folder 13.

89. *United States Statutes at Large* 62, pt. 1, 1275. This was designated Section 102 of the Housing Act of 1948, which became law on 10 August 1948.

90. "Charge 'Favoritism' On Steel for Homes," *New York Times,* 15 July 1948, 40. "Lustron Assured Steel for Houses," *Columbus Dispatch,* 5 August 1948. "That Lustron Affair," 93.

91. Ibid. "Help for Lustron," *Time* 52 (2 August 1948): 66.

92. "Lustron Head Charges 'Stall,'" *Columbus Dispatch,* 15 July 1948. "Big Plans for a Little House," *Business Week,* 16 October 1948, 44.

93. *United States Statutes at Large* 61, 526. Executive Order 9919, 13 *Federal Record* 59.

94. George F. Burley to Mathias Niewenhous, 6 April 1948 and 15 April 1948; and Niewenhous to Burley, 12 April 1948, Truman Papers, Official File 1930. Records of the Office of Industry Cooperation, General Records of the Department of Commerce, National Archives, Record Group 40.

95. Roy Ingersoll to John Steelman, 26 June 1948; Ingersoll to C. J. Rodman, 24 February 1950; and Ingersoll to Steelman, 27 February 1950, Truman Papers, Official File 1930.

96. Woods to Steelman, 4 December 1948, Truman Papers, Official File 1930.

97. Strandlund to Steelman, 10 May 1948, Truman Papers, Official File 1930.

98. Bartlett, "Lustron—The House That Lots of Jack Built," 68–71. "First Gas Furnace Started at Lustron," *Columbus Citizen,* 31 August 1948. "First Lustron Furnaces Are Fired," *Columbus Dispatch,* 31 August 1948. "Lustron Homes Begin," *New York Times,* 3 September 1948, 34.

Notes to Chapter 4

1. Strandlund to Emanuel Hogenson, 15 August 1949, CVA, Cabinet 1, Folder 15 contains the balcony anecdote and Strandlund's assessment of the reporter from *Architectural Forum.* "The Factory Built House Is Here, But Not the Answer to the $33 Million Question: How to Get It to Market?" *Architectural Forum* 90 (May 1949): 107.

2. William K. Wittausch, "Marketing Prefabricated Houses," *Harvard Business Review* 26 (November 1948): 693–94. James Marston Fitch, *American Building: The Forces That Shape It* (Boston: Houghton Mifflin, 1948), 342–58. Foster Gunnison, "The Economics of Mass-Distribution and Mass-Sales of Prefabricated Homes," *Prefabricated Homes* 2 (February 1944): 23. John Ely Burchard, "Prefabricated Housing and Its Marketing Problems," *The American Marketing Journal* (July 1935): 150–56. "Lustron Sales Program," LCR, Box 1, Folder 9. "Big Plans for a Little House," *Business Week,* 16 October 1948, 44. "Warmup at Lustron," *New York Times,* 3 September 1948, 34.

3. Wittausch, "Marketing Prefabricated Houses," 695–96.

4. Burnham Kelly, *The Prefabrication of Houses: A Study by the Albert Farwell Bemis Foundation of the Prefabrication Industry in the United States* (Cambridge, MA: Technology Press of the Massachusetts Institute of Technology, 1951), 92, 157–63, 391–94. George W. McKinney, Jr., "Residential Mortgage Lenders," *Journal of Finance* 7 (March 1952): 28–46. Williard Hull Brown, "The Sales Department Organization of the Lustron Corporation from Incorporation to Foreclosure," MBA thesis, Ohio State University, 1950, 13–18. Brown worked as a sales representative at Lustron during 1948 and 1949.

5. Kelly, *The Prefabrication of Houses,* 157–63. "Prefab Financing," *BusinessWeek,* 13 August 1949, 20. Wittausch, "Marketing Prefabricated Houses," 706–7. "Lustronics," *Newsweek* 34 (10 October 1949): 71.

6. Brown, "The Sales Department Organization of the Lustron Corporation," 18. Kelly, *The Prefabrication of Houses,* 166–68, 393–96.

7. Ibid., 703–4. Kelly, *The Prefabrication of Houses,* 87–88, 360–64, 372–82. Thomas S. Dicke, *Franchising in America: The Development of a Business Method, 1840–1980* (Chapel Hill: University of North Carolina Press, 1992), defines franchising as "a method of organization that combines large and small business into a single administrative unit . . . For the system to work, franchise holders, although legally independent, must conform to detailed standards of operation designed and enforced by the parent company" (pp. 2–3). Lustron's dealer network was an example of "product franchising" and comparable to the dealer relationship in the automotive industry. However, the marketing of prefabricated houses presented many challenges not found in the marketing of automobiles. Dealers could subvert the company's emphasis on "contextual" issues such as site and neighborhood planning. The issue of control over its dealer network thus became an important and ongoing concern.

8. "Lustron Sales Program," LCR, Box 1, Folder 9.

9. Wittausch, "Marketing Prefabricated Houses," 704.

10. "Lustron Profit Unit Plan," LCR, Box 1, Folder 5. Brown, "The Sales Department Organization of the Lustron Corporation," 16–17.

11. "Lustron Sales Program," LCR, Box 1, Folder 9. Brown, "The Sales Department Organization of the Lustron Corporation," 25, 27–29, 33.

12. Ibid.

13. U.S. Congress, Senate, Committee on Banking and Currency, *Problems of Independent Small Business: Lustron Dealers,* pt. 1, 81st Cong., 2d sess., 1 March 1950, 1–14. Brown, "The Sales Department Organization of the Lustron Corporation," 29–30, 35–36. "Facts About Lustron," LCR, Box 1, Folder 13. "Dealer Performance Report," LCR, Box 1, Folder 8. *Lustron Newsletter,* 3 September 1948. The plan to rely on 100 dealers in a thirty-seven-state area to market 30,000 house each year proved unrealistic but provides good evidence of the optimism of top management at this point in corporate history.

14. "Lustron Sales Program," LCR, Box 1, Folder 9. Brown, "The Sales Department Organization of the Lustron Corporation," 24, 35–36. "The Factory Built House Is Here," *Architectural Forum* 90 (May 1949): 112–13.

15. "Lustron Corporation Dealer Franchise Agreement," original document attached as Appendix D in Brown, "The Sales Department Organization of the Lustron Corporation."

16. "Lustron Sales Program," LCR, Box 1, Folder 9. "Dealer Performance Report," LCR, Box 1, Folder 8. Brown, "The Sales Department Organization of the Lustron Corporation," 23.

17. *Lustron Newsletter,* 19 March 1948.

18. "The Factory Built House Is Here," 112.

19. Ibid. *Lustron Newsletter,* 19 March 1948.

20. Ibid. "Lustron Sales Program," LCR, Box 1, Folder 9. "Dealer Performance Report," LCR, Box 1, Folder 9.

21. Brown, "The Sales Department Organization of the Lustron Corporation," 35–36, 46. "Lustron Sales Program," LCR, Box 1, Folder 9. "Lustron Dealers Get Construction

Training," *Columbus Dispatch,* 19 June 1948. "Facts about Lustron," LCR, Box 1, Folder 13.One technique devised by a dealer and promoted by the company was to scout out new subdivisions, purchase the best lots, and place a large sign on the lot reading "This Lot Certified for a Lustron Home."The tactic gave the company instant "fake prestige" by establishing the house in a traditional residential setting. "Everybody is delighted with the idea," one observer reported. Albert Farwell Bemis Foundation Records, Series III, Albert Farwell Bemis Research Project File, Box 5, Folder "Lustron Study Correspondence and Interviews, Institute Archives, and Special Collections, Massachusetts Institute of Technology, Cambridge, MA." Hereinafter cited as Bemis Foundation Records.

22. "Program, Lustron Dealer Convention, November 8 and 9, 1948, Columbus, Ohio," LCR, Box 2, Folder 3.

23. "Lustron Profit Unit Plan," LCR, Box 1, Folder 5. *Lustron Newsletter,* 13 August 1948.

24. Ibid.

25. Ibid. Kelly, *The Prefabrication of Houses,* 92. Brown, "The Sales Department Organization of the Lustron Corporation," 24.

26. "Lustron Profit Unit Plan," LCR, Box 1, Folder 5.

27. Ibid. "Lustron Sales Program," LCR, Box 1, Folder 9.

28. Ibid. "The Factory Built House Is Here," 112. *Architectural Forum* observed that the Lustron house would be well suited for installation on "some 30 million improved lots estimated to be standing vacant in U.S. cities."

29. "The Lustron Planning Guide," LCR, Box 2, Folder 3.

30. Ibid., introduction. Kenneth T. Jackson, *Crabgrass Frontier: The Suburbanization of the United States* (New York: Oxford University Press, 1985).

31. "The Lustron Planning Guide," 6.

32. Ibid., 7–8.

33. Ibid.

34. Ibid., 10–13.

35. Ibid.

36. Ibid., 14.

37. Ibid., 15–26.

38. Ibid., 27–30.

39. Ibid., 30–31.

40. Ibid., 32, 36–41.

41. Ibid., 42.

42. Ibid., 43–46.

43. Ibid., 47. Lustron turned to industrial designer Howard Ketchum, an expert on the relationship between color, perception, and behavior for advice on color schemes. As an engineer at DuPont during the 1930s, Ketchum advanced a "single color philosophy" that encouraged manufacturers to limit color selection. Ketchum had been involved with Chicago Vitreous's service station business as an advisor to several oil companies regarding color selection. For Lustron, Ketchum developed four color schemes: Desert Tan, Maize Yellow, Dove Grey, and Surf Blue. Complementary shades for each color adorned gable ends, door and window trim, and the roof. With the exception of the kitchen and utility room, the rest of the interior was gray. Ketchum believed that "yellow family" colors, which for him included cream, tan, and yellow-green shades, had an "energizing" effect and proved "conducive to conviviality, increased brain activity, and a sense of well

being." "Camera Tour through the Lustron Home," LCR, Box 2, Folder 2. Geoffrey T. Hellman, "Profiles," *New Yorker,* 8 March 1952, 41, 50.

44. Brown, "The Sales Department Organization of the Lustron Corporation," 42–43.

45. "Lustron Sales Program," LCR, Box 1, Folder 9. K. Reybold to Curt Mack, 4 January 1949, Truman Papers, Official File 1930. The Housing Act of 1948 expanded FHA insurance of private credit to prefabricators and other mass producers of housing. *United States Statutes at Large* 62, pt. 1, 1275–77.

46. Brown, "The Sales Department Organization of the Lustron Corporation," 54–55. "Lustron Sales Program," LCR, Box 1, Folder 9. "Facts about Lustron," LCR, Box 1, Folder 13.

47. "Facts about Lustron," LCR, Box 1, Folder 13. "Lustron Sales Program," LCR, Box 1, Folder 9. "Prefab Financing," *Business Week,* 13 August 1949, 20. Kelly, *The Prefabrication of Houses,* 395–96.

48. "Prefab Financing," *Business Week,* 13 August 1949, 20. "Facts about Lustron," LCR, Box 1, Folder 13.

49. Brown, "The Sales Department Organization of the Lustron Corporation," 49, 52–54.

50. Ibid., 55–56. "Lustron Sales Program," LCR, Box 1, Folder 9.

51. Ibid.

52. "What Holds Prefabs Back?" *Business Week,* 11 December 1948, 26. Mitchell, "What Ever Happened to Lustron Homes?" *APT Bulletin* 23 (1991): 47. *Congressional Record,* 5 February 1951, 1000–1001. Carl Strandlund to George Meader, 13 June 1950, LCR, Box 1, Folder 13. U.S. Congress, Senate, Committee on Banking and Currency, *Problems of Independent Small Business: Lustron Dealers,* pt. 1, 81st Cong., 2d sess., 1 March 1950, 15–16. U.S.; and pt. 2, 81st Cong., 2d sess., 14 March 1950, 89.

53. Major General Lemuel C. Shepherd, Jr. to Vice Admiral A. Stanton Merrill, 28 April 1949, LCR, Box 1, Folder 4.

54. Ibid.

55. William H. Whyte, *The Organization Man* (New York: Simon and Schuster, 1956). "Agreement between Lustron Corporation and Park Forest Homes, Inc., 3 August 1949," LCR, Box 1, Folder 2. "Lustron Closes Big Sales," *New York Times,* 12 November 1949, 20. Gregory C. Randall, *America's Original GI Town: Park Forest, Illinois* (Baltimore: Johns Hopkins University Press, 2000), 170–72.

56. "Agreement between Lustron Corporation and Park Forest Homes, Inc., 3 August 1949," LCR, Box 1, Folder 2. Randall, *America's Original GI Town,* 170–72.

57. Ibid.

58. Carl Strandlund to George Meader, 13 June 1950, LCR, Box 1, Folder 13. "Agreement between Lustron Corporation and the Strategic Air Command Housing Association, 29 June 1949," LCR, Box 1, Folder 2. Warren Bicknell, Jr. to Carl Strandlund, 21 November 1949, LCR, Box 1, Folder 2. *Problems of Independent Small Business: Lustron Dealers,* pt. 1, 28–31. U.S. Congress, Senate, Committee on Banking and Currency, *Problems of Independent Small Business: Lustron Dealers,* pt. 2, 81st Cong., 2d sess., 14 March 1950, 72–84.

59. *Lustron Newsletter,* 20 August 1948.

60. "The Factory Built House Is Here," 108–9. "Lustron Sales Program," LCR, Box 1, Folder 9.

61. "Lustron Corporation: Projected Operating Statements, 1949," LCR, Box 1, Folder

9. At the rate of production of forty houses per day, the company estimated a net profit of nearly $500,000 per month.

62. "Lustron Cost Sheet" and "Notes on Lustron Material," Bemis Foundation Records, Box 5, Folder "Lustron Housing." "Consumer Incomes: How They Moved Up," *Business Week,* 19 June 1948, 23. "Income Data: Planning Tool," *Business Week,* 17 May 1949, 94.

63. "Lustron Sales Program," LCR, Box 1, Folder 9. "The Factory Built House Is Here," 110. Tom Wolfe and Leonard Garfield, "'A New Standard for Living': The Lustron House, 1946–1950," in *Perspectives in Vernacular Architecture, vol. 3, ed.* Thomas Carter and Bernard Herman (Columbia: University of Missouri Press, 1989), 56–60. "First in New Ceramic Era: Lustron's $13 Million Plant," *Ceramics Industry* 51 (August 1948): 58.

64. "Lustron Comparative Operating Statement, March 1949," LCR, Box 1, Folder 9. "Facts about Lustron," LCR, Box 1, Folder 13.

65. "Lustron Lays Off 400," *New York Times,* 23 February 1949, 37. "12,000 Are Sent Home in Detroit—Lustron Lays Off 400," *New York Times,* 16 March 1949, 54.

66. "Lustron House Shipments," LCR, Box 1, Folder 2. "Record of Houses Loaded," LCR, Vertical File Materials (hereinafter cited as VFM). "Lustron Sales Program, LCR, Box 1, Folder 9. "Lustron Ships 24 Hours a Day," *New York Times,* 8 July 1949, 32. "Lustron Sales Record," *New York Times,* 1 August 1949, 25.

67. "Lustron Newsletter," 24 July 1949, LCR, Box 1, Folder 13.

68. The following hearings dealt in part with the issue of RFC aid to prefabricated housing firms: U.S. Congress, House of Representatives, Committee on Banking and Currency, *Housing Amendments of 1949,* 81st Cong., 1st sess., 25–29 July and 1–5, 8–9 August 1949; U.S. Congress, House of Representatives, Committee on Banking and Currency, *Housing Act of 1949,* 81st Cong., 1st sess., 7–8, 11–12, 25–28 April and 2–6, 9 May 1949; U.S. Congress, Senate, Committee on Banking and Currency, *Housing Amendments of 1949,* 81st Cong., 1st sess., 26–29 July 1949; U.S. Congress, Senate, Committee on Banking and Currency, *RFC Loan Policy,* 81st Cong., 1st sess., 22–30 June 1949; U.S. Congress, Senate, Committee on Banking and Currency, *Expansion of RFC Lending Policy,* 81st Cong., 1st sess., 2–18 August 1949.

69. *United States Statutes at Large* 62, pt. 1, 1275. U.S. Congress, Senate, Committee on Banking and Currency, *RFC Loan Policy,* 81st Cong., 1st sess., 22–30 June 1949, 1–3.

70. *RFC Loan Policy,* 1–10.

71. Ibid., 50–52.

72. Ibid., 83–84. "Bathtub Blues," *Time* 54 (4 July 1949): 55. "Lustron Gets RFC Loans," *New York Times,* 22 July 1949, 32.

73. "Strandlund Defends RFC Aid to Lustron," *New York Times,* 30 July 1949, 30. "To Study Lustron Loans," *New York Times,* 28 July 1949, 39.

74. "Lustron to Seek More RFC Money," *New York Times,* 4 August 1949, 37.

75. U.S. Congress, House of Representatives, Committee on Banking and Currency, *Housing Amendments of 1949,* 81st Cong., 1st sess., 25–29 July and 1–5, 8–9 August 1949, 425–40. "Lustron Losses Put at $14,754,000," *New York Times,* 5 August 1949, 30. "Loan of $14,500,000 Is Seen," *New York Times,* 16 August 1949, 36. "RFC Warns Lustron Gets All 'Prefab' Aid," *New York Times,* 1 September 1949, 42.

76. *Housing Amendments of 1949,* 416. "Lustron Losses Put at $14,754,000," *New York Times,* 5 August 1949, 30.

77. The authority to grant Lustron's original $15.5 million loan fell in part under Section 5 (d)(2) of the Reconstruction Finance Act, commonly referred to as the "National Defense Section."

78. *Housing Amendments of 1949,* 424, 441.

79. "Facts about Lustron," LCR, Box 1, Folder 13. "Lustron Gets RFC Loans," *New York Times,* 22 July 1949, 32.

80. "Lustron Lays Off 300 More," *New York Times,* 31 August 1949, 18.

81. "Bathtub Blues," *Time* 54 (4 July 1949): 55.

82. "That Lustron Affair," *Fortune* 40 (November 1949): 92–94. The White House continued to receive telegrams and letters criticizing the government's participation in Lustron. E. R. Murphy of Pasadena, CA asked Steelman if he was going to "waste more of U.S. money on a scheme with no hope of success?" E. J. Schell of Cincinnati, Ohio, referred to Lustron as a "hopeless proposition," and Henry D. Young, identifying himself as a Democrat, opined that "not one could be found in Lustron." Murphy to Steelman, 30 June 1949; Schell to Truman, 6 July 1949; and Young to Truman, 21 September 1949, Truman Papers, Official File 1930.

83. "To Study Lustron Loans," *New York Times,* 28 July 1949, 39. "RFC Warns Lustron Gets All 'Prefab' Aid," *New York Times,* 1 September 1949, 38. "Lustron Inquiry Asked," *New York Times,* 12 October 1949, 52.

84. "Lustron Corporation, Assembly Cost Exhibit," LCR, VFM 4107. A typical two-bedroom conventionally built house took approximately 1,600 man-hours to complete. Time sheets from several Lustron dealers in Minnesota, Illinois, and Indiana indicated an average installation time of 681 hours. "Lustron House Erection Time Reports," Bemis Foundation Records, Box 5, Folder "Lustron Housing."

85. Carl Koch and Andy Lewis, *At Home with Tomorrow* (New York: Rinehart, 1958), 110–12. *Who's Who in America* (Chicago: A. N. Marquis, 1955), 1733. "The Grandfather of Prefab: An Interview with Carl Koch," *Progressive Architecture* 75 (February 1994): 62. Williard Hull Brown, "The Sales Department Organization of the Lustron Corporation from Incorporation to Foreclosure," MBA thesis, Ohio State University, 1950, chapter 5, passim.

86. Koch and Lewis, *At Home with Tomorrow,* 113.

87. Ibid., 115–16.

88. "Lustron Sales Program," LCR, Box 1, Folder 9.

89. Koch and Lewis, *At Home with Tomorrow,* 116–17.

90. Ibid., 117–18.

91. Ibid., 118.

92. Ibid., 118–19. Burnham Kelly, *Design and the Production of Houses* (New York: McGraw-Hill, 1959), 108.

93. Koch and Lewis, *At Home with Tomorrow,* 102–21. See also "The Grandfather of Prefab."

94. Ibid., 122.

95. "Facts about Lustron," LCR, Box 1, Folder 13.

96. "Garage and Breezeway Variations with Lustron Homes," LCR, Box 2, Folder 1. W. A. Matheson, Sales Department, To All Dealers, "Package Price List of Models G-1 and G-2 Garage," 26 October 1949, LCR, Box 2, Folder 1.

97. "Camera Tour through the Lustron Home," LCR, Box 2, Folder 2. "'Living' Proof," LCR, Box 1, Folder 8. Brown, "The Sales Department Organization of the Lustron Corporation from Incorporation to Foreclosure," 83. "Lustron Adds Bedroom," *New York Times,* 25 December 1949, sec. 8, 2. The three-bedroom Westchester had 1,209 square feet. The three-bedroom Meadowbrook had 1,023 square feet, while the two-bedroom Meadowbrook had 775 square feet. The three-bedroom Newport had 961 square feet,

and the two-bedroom Newport had 713 square feet. Lustron also entered into an agreement with the Drexel Furniture Company and Lazarus Department Stores to offer "furniture packages" designed to suit each model. The pamphlet "Lustron Built It . . . Lazarus Furnished It" provided detailed listings of furniture and accessories for each room, along with prices "so easy to afford with Lazarus's 'Few Pennies a Day'" payment plan. "Lustron Built It . . . Lazarus Furnished It," LCR, Box 2, Folder 1.

98. "The Factory Built House Is Here," 108.

99. J. C. Tedlie, "Remarks," 14 September 1949, LCR, Box 1, Folder 5.

100. "The Lustron Home Zone and Price List," LCR, Box 1 Folder 8. "Home Zone" referred to sales territories, not to zoning regulations. "Daily House Erection Cost and Progress Report," LCR, Box 1 Folder 8.

101. "Lustron Sales Program," LCR, Box 1, Folder 9. Kelly, *The Prefabrication of Houses,* 371. "Bathtub Blues," *Time* 54 (4 July 1949): 55. Koch and Lewis, *At Home with Tomorrow,* 126.

102. "The Lustron Home," advertisement, LCR, Box 2, Folder 1.

103. "Lustron Sales Program," LCR, Box 1, Folder 9.

104. Ibid. Kelly, *The Prefabrication of Houses,* 88–89.

105. "The Factory Built House Is Here," 108. "*Forum* Yardstick Houses," *Architectural Forum* 86 (June 1947): 102–3. "Home Sweet Home—At a Price," *BusinessWeek,* 22 February 1947, 17.

106. Brown, "The Sales Department Organization of the Lustron Corporation from Incorporation to Foreclosure," 91–94.

107. "Notes on Conversation with Mr. Shire concerning D.C. Lustron Dealership," Bemis Foundation Records, Box 5, Folder "Lustron Housing." Tempers flared openly at a dealer conference in Columbus in May 1949. Lustron arranged a series of panels to permit open discussion of dealer–home office issues. So many dealers fired "hot, heavy, blunt, and embarrassing" questions that sales vice president Richard Jones turned out the lights and left the room.

108. "FHA Valuations and Commitments on the Lustron House," Bemis Foundation Records, Box 5, Folder "Lustron Housing." "The Factory Built House Is Here," 113. Kelly, *The Prefabrication of Houses,* 90–91, 388–89.

109. "What Holds Prefabs Back?" *Business Week,* 11 December 1948, 25–26. "Facts about Lustron," LCR, Box 1, Folder 13. Kelly, *The Prefabrication of Houses,* 90–92.

110. Ibid. "Building Codes under Fire," *BusinessWeek,* 23 March 1946, 41–52.

111. Ibid.

112. "The Factory Built House Is Here," 112. "Building Codes under Fire," 42.

113. "Big Plans for a Little House," *Business Week,* 16 October 1948, 44. "When Lustron Lost Its Luster," *Columbus Dispatch,* 23 July 1978.

114. Undated press clipping from the *Columbus Citizen,* LCR, VFM 4107.

115. "Prefab Makers Buckle Down," *BusinessWeek,* 23 August 1947, 18–19.

116. "When Lustron Lost Its Luster," *Columbus Dispatch,* 23 July 1978. Brown, "The Sales Department Organization of the Lustron Corporation from Incorporation to Foreclosure," 83–84. The weight of the trailer became an issue when the state of Tennessee refused to allow it to travel on its highways, effectively blocking Lustron from southwest markets. Dick Reedy successfully lobbied the Interstate Commerce Commission for an exemption, but the problem well illustrated both the figurative and literal roadblocks faced by the company as it wrestled with the transportation issue.

117. "Lustron Transportation Costs," Bemis Foundation Records, Box 5, Folder

"Lustron Study Correspondence and Interviews." Kelly, *The Prefabrication of Houses,* 404. "The Factory Built House Is Here," 110–11. "Lustron Sales Program," LCR, Box 1, Folder 9. See also U.S. Congress, Senate, Committee on Banking and Currency, *Study of Reconstruction Finance Corporation: Lustron Corp.—Transportation Contract,* 81st Cong., 2d sess., 26–30 June 1950.

118. Prefabricated House Manufacturers' Institute, *Washington News Letter,* 8 October 1948, 3, quoted in Kelly, *The Prefabrication of Houses,* 397. Brown, "The Sales Department Organization of the Lustron Corporation from Incorporation to Foreclosure," 91–92.

119. Kelly, *The Prefabrication of Houses,* 92–94, 396–97. "What's Stalling Lustron," *Business Week,* 29 October 1949, 25.

120. "The Factory Built House Is Here," 108, 112. Kelly, *The Prefabrication of Houses,* 395–96. "Bathtub Blues," *Time* 54 (4 July 1949): 55.

121. U.S. Congress, Senate, Committee on Banking and Currency, *Problems of Independent Small Business: Lustron Dealers,* 81st Cong., 2d sess., 1 and 14 March 1950, passim.

122. "Lustron Directors Quit," *New York Times,* 15 October 1949, 26. Brown, "The Sales Department Organization of the Lustron Corporation from Incorporation to Foreclosure," 84, 93–96.

123. "Lustron Sales Program," LCR, Box 1, Folder 9. Brown, "The Sales Department Organization of the Lustron Corporation from Incorporation to Foreclosure," 73–74.

124. "Dealer Performance Report: 1949," LCR, Box 1, Folder 5.

125. "Lustron Sales Program," LCR, Box 1, Folder 9. Brown, "The Sales Department Organization of the Lustron Corporation from Incorporation to Foreclosure," 98–101.

126. Ibid. Although Lustron had little choice at this point, there were dangers in relying too much on large-volume sales. Unless project orders flowed steadily, they placed a tremendous, but temporary strain on operations. A "slack period" typically followed the completion of a large order. Ironically, such a pattern was characteristic of conventional contracting businesses. But, unlike the contractor, a mass-production prefabricated housing firm retained constant overhead expenses such as plant investment and labor costs. Ideally, many small sales based on a large dealer network best suited large-scale factory-made housing because it diversified the sales risk and rationalized continuous production.

127. "Lustron Sales Program," LCR, Box 1, Folder 9.

Notes to Chapter 5

1. The RFC divided its loans into five categories: industrial and commercial enterprises; railroads; financial institutions; political subdivisions of states and territories; and catastrophe loans. The agency reported $384,345,549 in outstanding loans to industrial and commercial enterprises in 1949. Beryl Wayne Sprinkel, "Economic Consequences of the Operations of the Reconstruction Finance Corporation," *The Journal of Business of the University of Chicago* 25 (October 1952): 211–24.

2. "Facts about Lustron," LCR, Box 1, Folder 13. "Lustron Fleet Sales Projects," LCR, Box 1, Folder 9. Fleet and military sales looked especially promising, with orders totaling nearly 13,000 units.

3. See U.S. Congress, Senate, Committee on Banking and Currency, *Employment of RFC Personnel,* 81st Cong., 1st sess., 16 June 1949.

4. The position of Federal Loan Administrator was held by Jesse Jones, the popular and influential chairman of the RFC 1932–1945 and secretary of commerce under Franklin Roosevelt. When Roosevelt removed Jones in 1945 from all his positions, the RFC began a downward spiral at the hands of less able leadership. Further, the RFC lost the support and confidence of Congress, which had rarely questioned its administration during Jones's tenure. Andrew J. Dunar, *The Truman Scandals and the Politics of Morality* (Columbia: University of Missouri Press, 1984), 78–79. Marshall E. Dimock, "Government Corporations: A Focus of Policy and Administration, I," *The American Political Science Review* 43 (October 1949): 899–921.

5. U.S. Congress, Senate, Committee on Banking and Currency, *Study of Reconstruction Finance Corporation—Favoritism and Influence,* S. Rept. 76, 82d Cong., 1st sess., 5 February 1951, 3.

6. Robert H. Ferrell, *Harry S. Truman: A Life* (Columbia: University of Missouri Press, 1994), 363; Dunar, *The Truman Scandals and the Politics of Morality,* 81; Haynes Johnson and Bernard Gwertzman, *Fulbright: The Dissenter* (Garden City, NY: Doubleday, 1968), 100–105.

7. U.S. Congress, Senate, Committee on Banking and Currency, *Study of Reconstruction Finance Corporation—Lending Policy,* pt. 2, 82d Cong., 1st sess., 21–23, 26–27 February and 1 March 1951, 933.

8. F. E. Brickel, assistant manager of the Loan Agency of the Reconstruction Finance Corporation, to D. W. Boylan, secretary of Lustron, 13 July 1948, in *Study of Reconstruction Finance Corporation—Lending Policy,* pt. 2, 934.

9. U.S. Congress, Senate, Committee on Banking and Currency, *Study of Reconstruction Finance Corporation and Proposed Amendment of RFC Act,* S. Rept. 649, 82d Cong., 1st sess., 20 August 1951, 44–45. *Study of Reconstruction Finance Corporation—Favoritism and Influence,* 6–7.

10. U.S. Congress, Senate, Committee on Banking and Currency, *Study of Reconstruction Finance Corporation—Lending Policy,* pt. 2, 946–48. Jules Abels, *The Truman Scandals* (Chicago: H. Regnery Co., 1956), 86–87. Donald Dawson later served as a personnel assistant to Truman in the White House. His duties included "vetting" potential nominees for administration posts, a position that critics charged he used to appoint "cronies" to influential positions. The question remains whether Dawson used his influence to get Young the Lustron job, a charge that the Fulbright committee raised, since Young had no experience in the housing industry.

11. *Study of Reconstruction Finance Corporation—Lending Policy,* pt. 2, 938.

12. Edward Brown, *Democracy in Action* ([Washington, DC], 1948), n.p.

13. *Study of Reconstruction Finance Corporation and Proposed Amendment of RFC Act,* 47.

14. Ibid., 46. *Study of Reconstruction Finance Corporation—Lending Policy,* pt. 2, 934.

15. *Study of Reconstruction Finance Corporation—Lending Policy,* pt. 2, 707, 945, 1013, 1039. *Study of Reconstruction Finance Corporation and Proposed Amendment of RFC Act,* 46–47.

16. *Study of Reconstruction Finance Corporation—Lending Policy,* pt. 2, 1318.

17. Ibid., 934.

18. Ibid., 934.

19. Ibid., 934–35.

20. Ibid., 935. *Study of Reconstruction Finance Corporation and Proposed Amendment of RFC Act,* 46.

21. *Study of Reconstruction Finance Corporation—Lending Policy,* pt. 2, 935–36.

22. Ibid., 936.

23. Ibid., 965. The Jacobs-Hunt report also recommended that Young and "others" be given the opportunity to purchase enough Lustron stock to "offset the 80,000 shares that Mr. Strand[lund] and Mrs. Strand[lund] own at a price not to exceed *.01 cents per share*" (emphasis mine).

24. Ibid., 937.

25. Harley Hise to Carl Strandlund, 4 October 1949, LCR, Box 1, Folder 4.

26. "Lustronics," *Newsweek* 34 (10 October 1949): 71.

27. *Study of Reconstruction Finance Corporation—Lending Policy,* pt. 2, 938.

28. H. Res. 393, *Congressional Record* 95, 974.

29. *Congressional Record* 95, A6206.

30. Ibid., A6181. Vursell also introduced into the record a transcript of a broadcast made by Fulton Lewis over the Mutual Radio Network on September 29. Lewis, an old enemy of Wyatt and consistent Lustron critic, savaged the company and the RFC and ended by noting that disadvantaged Americans could get all the government aid they desired by going into the manufacture of prefabricated houses.

31. "Lustron Inquiry Asked," *New York Times,* 12 October 1949, 52.

32. *Congressional Record* 95, 14457–58.

33. Ibid., 14459–61.

34. "What's Stalling Lustron?" *Business Week,* 29 October 1949, 25. "Lustron in Arrears, R.F.C. Loans Halted," *New York Times,* 29 December 1949, 39.

35. *New York Times,* 12 November 1949, 20. "What's Stalling Lustron?" *Business Week,* 29 October 1949, 25. "Whither Lustron?" *Newsweek* 35 (27 February 1950): 60. "Local Firm Uses No RFC Funds for Lustron Homes," *Columbus Dispatch,* n.d., LCR, VFM 4107.

36. *Study of Reconstruction Finance Corporation—Lending Policy,* pt. 2, 938.

37. Ibid.

38. Ibid., 939.

39. Ibid.

40. "Lustron Warned on Debt Owed R.F.C.," *New York Times,* 30 December 1949, 31. Harvey Gunderson, Strandlund's best ally among the RFC Board, told the *New York Times* that the RFC would "have to do something very soon" unless the company submitted "a plan to make itself solvent."

41. Carl Strandlund to Richard C. Dyas, 6 January 1950, LCR, Box 1, Folder 4. "Lustron Plan Is Filed," *New York Times,* 7 January 1950. Phillip Klutznick was the developer of Park Forest, Illinois, and had been a close friend of Strandlund's since the two negotiated a contract to use Lustron houses in Park Forest in January 1949. Gregory C. Randall, *America's Original GI Town: Park Forest, Illinois* (Baltimore: Johns Hopkins University Press, 2000), 172.

42. Ibid. "Lustron Fleet Sales Projects," LCR, Box 1, Folder 9. "Lustron Profits Hinge on New Aid," *Columbus Dispatch,* 9 January 1950.

43. Carl Strandlund to Richard C. Dyas, 6 January 1950, LCR, Box 1, Folder 4.

44. Ibid.

45. *Study of Reconstruction Finance Corporation—Lending Policy,* pt. 2, 939. "RFC to Refuse Lustron's Plea," *Columbus Dispatch,* 10 January 1950. "RFC Rejects Lustron's Plan for Reorganizing," *Columbus Dispatch,* 11 January 1950. "R.F.C. Bars Reorganizing As Proposed by Lustron," *New York Times,* 11 January 1950, 35.

46. Quoted in Irving Welfeld, *Where We Live: A Social History of American Housing* (New York: Simon and Schuster, 1988), 29. "Extract from and Comments on the Booz, Allen, and Hamilton Report to the RFC on Lustron," Bemis Foundation Records, Box 5, Folder "Lustron Housing."

47. "Hot Spot for Lustron," *Newsweek* 35 (23 January 1950): 60.

48. "Senate Committee Stalls RFC Action against Lustron," *Columbus Dispatch,* 11 January 1950.

49. "Lustron Still Has Time to Placate RFC," *Columbus Dispatch,* 12 January 1950. "Lustron President in Washington for Parley With RFC," *Columbus Dispatch,* 13 January 1950.

50. "Lustron Has New Setback," *Columbus Dispatch,* 16 January 1950. "RFC Preparing Foreclosure Suit against Lustron," *Columbus Dispatch,* 18 January 1950. "U.S. to Act on Lustron," *New York Times,* 18 January 1950, 46.

51. "Lustron Has New Setback."

52. "Banks Anticipate Foreclosure Action against Lustron," n.d., LCR, VFM 4107.

53. "Suit Fails to Halt Lustron Production," n.d., LCR, VFM 4107.

54. Valentine wrote to Steelman expressing "the faithful belief of all the people employed by the Lustron Company, that [if] they be permitted the chance to produce, the indebtedness to the government will be paid in full . . . [and] the welfare of untold thousands will be assured in the future." Valentine to Steelman, 17 February 1950, Truman Papers, Official File 1930. "Lustron Battles U.S. Foreclosure," *New York Times,* 21 February 1950, 33.

55. "R.F.C. Orders Foreclosure Action Undertaken against Lustron Corp," *New York Times,* 15 February 1950, 37. "U.S. Orders Court Action on Lustron," *Columbus Dispatch,* 17 January 1950. Carl Strandlund to Richard C. Dyas, 6 January 1950, LCR, Box 1, Folder 4.

56. "U.S. to Study Using Big Lustron Plant," *New York Times,* 16 February 1950, 37. "RFC Orders Foreclosure of Lustron's Loan," n.d., LCR, VFM 4107.

57. "Government Files Suit to Foreclose on Lustron," *New York Times,* 22 February 1950, 41. "RFC Cuts Off Aid, Forecloses on Lustron," *Engineering News-Record* 144 (23 February 1950): 27. "Lustron Goes on the Block," *Business Week,* 25 February 1950, 24. "Whither Lustron?" *Newsweek* 35 (27 February 1950): 60. "RFC Will Press Action on Lustron, *New York Times,* 26 February 1950, 48. "Lustron Action Sought," *New York Times,* 28 February 1950, 43. *Study of Reconstruction Finance Corporation—Lending Policy,* pt. 2, 940.

58. "Lustron Newsletter," 24 February 1950, LCR, Box 1, Folder 7. "Dealers to Aid Lustron," *New York Times,* 1 March 1950, 47. "Lustron Dealers Enter Battle," n.d., LCR, VFM 4107.

59. U.S. Congress, Senate, Committee on Banking and Currency, *Problems of Independent Small Business—Lustron Dealers,* 81st Cong., 2d sess., 1 March 1950, 1–3.

60. Ibid., 4. See also Lobdell to Truman, 2 March 1950, Truman Papers, Official File 1930.

61. *Problems of Independent Small Business—Lustron Dealers,* 5–6.

62. Ibid., 7–10.

63. Ibid., 6.

64. Ibid., 13.

65. Ibid., 20–26.

66. Ibid., 29–31.

67. *In Re Lustron Corp. Reconstruction Finance Corp. v. Lustron Corp.,* 184 F.2d 789. "Receiver Is Appointed for Lustron Corp. for Operation of Plant Another 30 Days," *New York Times,* 7 March 1950, 35.

68. "Lustron Officials Relieved of Posts," *New York Times,* 9 March 1950, 45. The firings cut Lustron's payroll from $156,000 to $70,000 and gave Foraker additional flexibility to continue production.

69. W. V. Burnell to R. Harold Denton, 21 August 1951, LCR, Box 1, Folder 4.

70. "Lustron Gets Extension, Indefinite Period Set by Court for Receivership Operation," *New York Times,* 4 April 1950, 50. Carl Strandlund to George Meader, 13 June 1950, LCR, Box 1, Folder 13. Most of Lustron's critics in Congress praised the RFC foreclosure action. Representative Robert Rich (R-PA) said, "no matter what will be lost through foreclosure proceedings, one rathole down which the taxpayers' money has been slipping will be plugged." Cited in *Congressional Record* 96, 16 February 1950, A1157. See also the remarks of Representative Albert Cole in *Congressional Record* 96, 16 March 1950, A1994.

71. "Lustron Newsletter," 24 February 1950, LCR, Box 1, Folder 7. "End of Lustron Near," *New York Times,* 26 April 1950, 51. "Dealers to Aid Lustron," *New York Times,* 1 March 1950, 47. "Lustron Goes on the Block," *Business Week,* 25 February 1950, 24. "A Close Call for Fruehauf," *Business Week,* 7 April 1951, 72–77.

72. The raw material inventory was excluded from the RFC's bid because it was not included in a lien on the company's assets claimed by the agency following Lustron's default on the loan agreements. "R.F.C. Bids $6,000,000 for the Lustron Plant," *New York Times,* 7 June 1950, 43. "Lustron Goes on the Block," *Business Week,* 25 February 1950, 24.

73. "Three Lustron Dealers Block Plant Sale," *New York Times,* 9 June 1950, 36. Judge Sullivan sat on the U.S. District Court for the Northern District of Illinois, Eastern Division.

74. "R.F.C. Pushes Sale of Lustron's Assets," *New York Times,* 13 June 1950, 41. "Two Courts Conflict over Lustron Case," *New York Times,* 14 June 1950, 47.

75. "R.F.C. Gets Lustron Plant Despite Conflicting Order," *New York Times,* 17 June 1950, 18. "Inquiry on Lustron Asked by U.S. Court," *New York Times,* 15 June 1950, 49.

76. "Inquiry on Lustron Asked by U.S. Court," 49.

77. "Justice Department Agrees to Investigate Lustron Corp. without Unnecessary Delay," *New York Times,* 20 June 1950, 40.

78. "Lustron Is Headed for Appeals Court," *New York Times,* 21 June 1950, 41. "Action on Lustron Enters New Phase," *New York Times,* 28 June 1950, 43.

79. U.S. Congress, Senate, Committee on Banking and Currency, *Study of Reconstruction Finance Corporation: Lustron Corp.—Transportation Contract,* 81st Cong., 2d sess., 26–30 June 1950, 5–6. "Lustron Loss Put at $500,000 in Deal," *New York Times,* 27 June 1950, 38. *Congressional Record* 97, 23 August 1951, A5348–49.

80. *Study of Reconstruction Finance Corporation: Lustron Corp.—Transportation Contract,* 7–10. In a letter to Truman, Representative Cole detailed his perceptions of the Lustron–Commercial Home situation and criticized the administration for a lack of oversight. "Mr. President," Cole concluded, "I respectfully call your attention to the situation in the hope that experiments in socialistic holding companies be not authorized in the guise of commercial activity." Cole to Truman, 21 August 1950, Truman Papers, Official File 1930. Cole also praised the investigation in the press: "It looks to me . . . as though the R.F.C. let somebody get away with something right under their noses." Quoted in "R.F.C. Again Attacked on Loans to Lustron," *New York Times,* 9 August 1950, 44. "Senators Study $334,000 'Loan' to Officials of Concern Accused of Defrauding Lustron," *New York Times,* 30 June 1950, 38.

81. *Congressional Record* 96, 11 August 1950, 12281. "Lustron Deal Scorned," *New York Times,* 12 August 1950, 20. An immediate result of the inquiry was the resignation of Harley Hise as chairman of the RFC. In his letter of resignation, Hise blamed Congress for the "Lustron mess" and reminded his critics that the RFC was "a lending and not a police agency." Quoted in "Hise, Ousted R.F.C. Head, Blames Congress for Loans to Lustron,"

New York Times, 18 August 1950, 29. In April 1952 a federal grand jury indicted Commercial Home and three of its officials on fraud charges. All defendants were subsequently acquitted. "Truck Rent Fraud on Lustron Charged," *New York Times,* 30 April 1952, 46. "Defunct Firm Acquitted," *New York Times,* 9 October 1953, 43. In late August, General Electric "expressed an interest" in purchasing the Lustron plant to produce aircraft engines. However, the legal impasse over the foreclosure caused top G.E. executives to shy away and the company quickly withdrew its interest in the plant. "G.E. Is Now 'Interested' in Idle Lustron Factory," *New York Times,* 29 August 1950, 37.

82. "Receiver Seeks Pay For Lustron Staff," *New York Times,* 18 July 1950, 36. "U.S. Court of Appeals Sets Aug 1 Hearing To Straighten Out Lustron Legal Tangle," *New York Times,* 23 July 1950, sec. 3, 2. "Appeals Bench Gets Lustron Court Row," *New York Times,* 30 September 1950, 20.

83. *In re Lustron Corp. Reconstruction Finance Corp. v. Lustron Corp.* 184 F.2d 789, 23 October 1950. *In re Lustron Corp. Lafayette Steel Co. v. Lustron Corp.* 184 F.2d 798, 23 October 1950.

84. *United States Statutes at Large* 64, pt. 1, 1279.

85. Andrew J. Dunar, *The Truman Scandals and the Politics of Morality* (Columbia: University of Missouri Press, 1984), 85–86.

86. *Congressional Record* 97, 5 February 1951, 999–1004.

87. Carl Strandlund to W. Elmer Harber, 28 December 1950, LCR, Box 1, Folder 4.

88. Ibid.

89. *Congressional Record* 97, 12 January 1951, 194–95.

90. *Congressional Record* 97, 2 February 1951, A535. See also Sutton to Truman, 31 January 1951, Truman Papers, Official File 1930.

91. *Congressional Record* 97, 2 February 1951, A535.

92. U.S. Congress, Senate, Committee on Banking and Currency, *Proposed Disposition of Lustron Plant,* 82d Cong., 1st sess., 13 February 1951, 8–9. Unions representing Lustron employees sent many telegrams and letters to Sparkman and to the White House urging preservation of the plant for housing production. See, for example, T. E. Waller to Sparkman, 16 February 1951 and T. J. Conroy to Steelman, 15 February 1951, Truman Papers, Official File 1930.

93. Ibid., 51.

94. Ibid., 52. "Navy Gets Lustron Plant," *New York Times,* 30 March 1951, 13.

95. *Reconstruction Finance Corp., Petitioner, v. Lustron Corp. et al.* 340 U.S. 946, 12 March 1951. *LaFayette Steel Co., Inc., Petitioner, v. Lustron Corp. et al.* 340 U.S. 946, 12 March 1951. "High Court Denies Lustron Review," *New York Times,* 13 March 1951, 49.

96. "Auctioneer Knocks Down Remnants of Lustron Dream," *Business Week,* 21 July 1951, 23. "Lustron Assets Studied," *New York Times,* 16 June 1951. A motel construction firm bought many of the remaining porcelain-enameled steel panels. The garages behind the "Top of the Mark" roadside motel near Canton, Ohio, constructed in 1953, remain as perhaps the only verified example of the commercial application of material from Lustron's bankruptcy sale.

97. Dunar, *The Truman Scandals and the Politics of Morality,* 90–95. Alonzo L. Hamby, *Man of the People: A Life of Harry S. Truman* (New York: Oxford University Press, 1995), 586–87. Douglas Knerr, "The Reconstruction Finance Company Is Created," in *Great Events from History II: Business and Commerce* (Los Angeles: Salem Press, 1994), 632.

98. "R.F.C. Drops Action over Lustron Loan," *New York Times,* 16 January 1952, 15. On April 12, 1954 federal bankruptcy referee Wallace Streeter awarded $2 million to the RFC from Lustron's estate. Streeter also awarded ninety-five other Lustron creditors

a total of $342,000, an average of 25 percent of their original claims. "R.F.C. Gets $2,000,000 from Lustron Assets," *New York Times,* 13 April 1954, 55.

99. "Heads Jerry O'Mahony, Manufacturer of Diners," *New York Times,* 2 September 1953, 40. *Minneapolis Tribune,* 12 September 1982.

Notes to Chapter 6

1. Kim Keister, "Showing Its Metal," *Historic Preservation* 47 (January–February 1995): 36–38. Bonnie McGrath, "Still Going . . . All Steel Lustrons Have Become a Permanent Blip in Housing Industry," *Chicago Tribune,* 28 October 1995, 4. Tim Zorn, "Steel House of the 50s Open For Tours," *The Fresno Bee,* 22 May 1994, E10.

2. Keister, "Showing Its Metal," 38, 96.

3. The use of the term *evangelist* to describe a developer of strong customer loyalty to a product was popularized by the computer industry.

4. "Home Owner Speaks," *Business Week,* 25 February 1950, 64.

5. Ibid.

6. "A Study on Lustron Homes Made among 200 Lustron Home Owners," LCR, Box 1, Folder 13.

7. Ibid., 3–4, 7. "Findings from Lustron Survey—August 1953," LCR, Box 1, Folder 13.

8. Ibid.

9. "A Study on Lustron Homes," 4–5.

10. Ibid., 6.

11. Ibid., 6–7.

12. Thomas T. Fetters and Vincent Kohler, *The Lustron Home: A History of a Postwar Prefabricated Housing Experiment* (Jefferson, NC: McFarland, 2002), 122, 126.

13. "Porcelain Homes Stand Test of Time," *Chicago Sun Times,* 5 February 1996, 8.

14. Charles Stanley, "Rare Prefab in Path of Destruction," *Chicago Tribune,* 24 January 1996, Metro Southwest Section, 1. See also "Relocation of a Lustron House" at http://www.fema.gov/hp/lustron.shtm.

15. Bill Dunn, "History's a Game for Middleton," *Capital Times,* 1 February 1996, 4.

16. http://www.leonardomusic.com/Lustronpage.htm.

17. Thomas S. Dicke, *Franchising in America: The Development of a Business Method, 1840–1980* (Chapel Hill: University of North Carolina Press, 1992), 8–9.

18. Russell Davenport, "A *Life* Round Table on Housing," *Life* (31 January 1949): 74.

19. *United States Statutes at Large* 63, pt. 1, 413.

20. http://www.gsd.harvard.edu/studio/s97/burns/intro.html.

21. Mike McClintock, "Steel Frame Houses: Looking at an Alternative to Lumber," *The Washington Post,* 15 February 1996, T7. "Home Builders Eye Steel for Support," *Bangor Daily News,* 25 December 1995, 12. Rich Davis, "Steel Housing: An Old Ideal That's New Again," *Chicago Tribune,* 21 March 1993, N2. Several firms cited the use of steel framing for houses in earthquake-prone areas as an initial target market.

22. See the Steel Framing Alliance's website at http://www.steelframingalliance.com. Steve Farr, "Steelmakers See Opportunity in Housing," *Los Angeles Times,* 26 January 1996, D7.

23. "Why Steel: Because Steel Builds Better Houses," http://www.steelframingalliance.com/why/index.html.

24. U.S. Congress, Senate, Committee on Banking, Housing, and Urban Affairs, *Manufactured Housing Provisions of H.R. 5157,* 101st Cong., 2d sess., 26 July 1990, 1–5. Title IX of the Cranston-Gonzalez National Affordable Housing Act (Public Law 101–625) created the National Commission on Manufactured Housing. 104 Stat. 4079.

25. U.S. Congress, House of Representatives, Committee on Banking, Finance, and Urban Affairs, *Manufactured Housing,* 102d Cong., 2d sess., 29 September 1992, 1–15.

26. 108 Stat. 4645.

27. Ruth Knack, "House-in-a-Box," *Planning* 61 (August 1995): 10–12.

28. Ibid.

29. See the PATH website at http://www.pathnet.org/.

30. "Factory Built Components and Housing Systems," Conference Report of the Centre for Advanced Wood Processing, 18 and 19 March 2002.

31. "Meeting Our Nation's Housing Challenges," Report of the Bipartisan Millennial Housing Commission Appointed by the Congress of the United States (Washington, 2002), 72–73.

32. "Porcelain Enamel Homes Have 27-Year History," 30 September 1959, LCR, Box 1, Folder 7, 2–3.

INDEX

Urban Life and Urban Landscape Series
Zane L. Miller, General Editor

The series examines the history of urban life and the development of the urban landscape through works that place social, economic, and political issues in the intellectual and cultural context of their times.

Cincinnati, Queen City of the West, 1819–1838
Daniel Aaron

Domesticating the Streets: The Reform of Public Space in Hartford, 1850–1930
Peter C. Baldwin

Proportional Representation and Election Reform in Ohio
Kathleen L. Barber

Fragments of Cities: The New American Downtowns and Neighborhoods
Larry Bennett

The Lost Dream: Businessmen and City Planning on the Pacific Coast, 1890–1920
Mansel G. Blackford

Merchant of Illusion: James Rouse, American's Salesman of the Businessman's Utopia
Nicholas Dagen Bloom

Suburban Alchemy: 1960s New Towns and the Transformation of the American Dream
Nicholas Dagen Bloom

Planning for the Private Interest: Land Use Controls and Residential Patterns in Columbus, Ohio, 1900–1970
Patricia Burgess

Cincinnati Observed: Architecture and History
John Clubbe

Lancaster, Ohio, 1800–2000: Frontier Town to Edge City
David R. Contosta

Suburb in the City: Chestnut Hill, Philadelphia, 1850–1990
David R. Contosta

Main Street Blues: The Decline of Small-Town America
Richard O. Davies

For the City as a Whole: Planning, Politics, and the Public Interest in Dallas, Texas, 1900–1965
Robert B. Fairbanks

Making Sense of the City: Local Government, Civic Culture, and Community Life in Urban America
Edited by Robert B. Fairbanks and Patricia Mooney-Melvin

The Mysteries of the Great City: The Politics of Urban Design, 1877–1937
John D. Fairfield

Faith and Action: A History of the Catholic Archdiocese of Cincinnati, 1821–1996
Roger Fortin

Cincinnati in 1840: The Social and Functional Organization of an Urban Community during the Pre–Civil War Period
Walter Stix Glazer

The Poetics of Cities: Designing Neighborhoods That Work
Mike Greenberg

History in Urban Places: The Historic Districts of the United States
David Hamer

The Failure of Planning: Permitting Sprawl in San Diego Suburbs, 1970–1999
Richard Hogan

Columbus, Ohio: A Personal Geography
Henry L. Hunker

Getting around Brown: *Desegregation, Development, and the Columbus Public Schools*
Gregory S. Jacobs

Regionalism and Reform: Art and Class Formation in Antebellum Cincinnati
Wendy Katz

Building Chicago: Suburban Developers and the Creation of a Divided Metropolis
Ann Durkin Keating

Silent City on a Hill: Landscapes of Memory and Boston's Mount Auburn Cemetery
Blanche Linden-Ward

Plague of Strangers: Social Groups and the Origins of City Services in Cincinnati, 1819–1870
Alan I. Marcus

Visions of Place: The City, Neighborhoods, Suburbs, and Cincinnati's Clifton, 1850–2000
Zane L. Miller

Boss Cox's Cincinnati: Urban Politics in the Progressive Era
Zane L. Miller

Changing Plans for America's Inner Cities: Cincinnati's Over-the-Rhine and Twentieth-Century Urbanism
Zane L. Miller and Bruce Tucker

Polish Immigrants and Industrial Chicago: Workers on the South Side, 1880–1922
Dominic A. Pacyga

The Rise of the City, 1878–1898
Arthur Meier Schlesinger

The New York Approach: Robert Moses, Urban Liberals, and Redevelopment of the Inner City
Joel Schwartz

Designing Modern America: The Regional Planning Association and Its Members
Edward K. Spann

Hopedale: From Commune to Company Town, 1840–1920
Edward K. Spann

Visions of Eden: Environmentalism, Urban Planning, and City Building in St. Petersburg, Florida, 1900–1995
R. Bruce Stephenson

Welcome to Heights High: The Crippling Politics of Restructuring America's Public Schools
Diana Tittle

Washing "The Great Unwashed": Public Baths in Urban America, 1840–1920
Marilyn Thornton Williams